AF379531

EVOLUTION EVOLVING

What Everyone Needs to Know

Patricia A. Williams

Copyright © 2009 by Patricia A. Williams

ISBN 0-7414-5580-3

Published by:

INFI∞ITY
PUBLISHING.COM
1094 New DeHaven Street, Suite 100
West Conshohocken, PA 19428-2713
Info@buybooksontheweb.com
www.buybooksontheweb.com
Toll-free (877) BUY BOOK
Local Phone (610) 941-9999
Fax (610) 941-9959

Printed in the United States of America

Published July 2010

The alternative to thinking in evolutionary terms is not
to think at all.

Sir Peter Medawar

TABLE OF CONTENTS

INTRODUCTION

As part of my graduate coursework in philosophy, I read Charles Darwin's *On the Origin of Species by Means of Natural Selection* and fell in love. "This," I said to myself, "is what I want to do with the rest of my life." There is much to love. The book is a masterpiece of philosophical argument. It is also clearly written, filled with fascinating facts and evident affection for animals. It is also logical—a philosopher's delight. Moreover, Darwin so well understood his own theory that contemporary philosophers and biologists working in the field of evolutionary biology often reread it annually for fresh understandings. Modern biology is evolutionary biology, and evolutionary biology today still derives many of its fundamental insights from Darwin.

Studies of evolution include both fact and theory. It is a fact that groups of organisms evolve, that is, they change over time. The evidence for the fact of evolution is everywhere. Many people think of the fossils that show the history of life on Earth. However, Darwin discovered evidence for evolution from his observations of animals living on islands. They represented different species, yet they all seemed related. He concluded they were related, that they evolved on the islands. Today perhaps the best evidence comes from genes. All organisms—bacteria, yeast, insects, reptiles, birds, and mammals (including us)—share some genes. About 3.8 billion years ago, a population of organisms lived that carried these genes. Every living thing possesses them.

Evolution is also a theory about causes. The main cause of evolution is natural selection. Natural selection is a force in nature that preserves some organisms and destroys others. Natural selection may be environmental, like drought that kills all the fish in a disappearing lake. It may also consist of interactions among species, like that between cat and mouse. Or it can occur within a single species, like mate choice by peahens that gave peacocks their famous tails. Natural selection is also well-supported with evidence. The evidence comes from experiments with living organisms in laboratories and from observations of nature.

This book is about major discoveries in biology and their impact on our understanding of evolution since Darwin published *On the Origin of Species* in 1859. At the same time, it is also about evolution as science, about whether it is good science, whether its new discoveries constitute good science, and whether they support Darwin's original concepts.

Darwin's work on evolution produced four major accomplishments. First, it provided a mechanism (natural selection) that explains how evolution occurs. Second, it united in one synthesis information from fields of biology that seemed separate. Third, it offered testable predictions. Fourth, it inspired new research in every field it touched, creating new publications, techniques, and institutions. It continues to inspire research today.

Most new discoveries after Darwin tended to support Darwin's work. One, however, threatened to undo his synthesis. This was the rediscovery in 1900 of Mendel's work on genes that proved inheritance discontinuous, whereas Darwin's concept of gradual change seemed to imply that inheritance is continuous. However, biologists successfully combined Mendel's discoveries with Darwin's gradual evolution and declared Darwin's synthesis restored under the name New Synthesis (or neo-Darwinism) in 1947. Nonetheless, the New Synthesis remained incomplete. A third synthesis called sociobiology, begun in 1964, restored two other elements from Darwin's original synthesis—the importance to evolution of relatives and of sexual selection. Now only developmental biology remains excluded. As of the 1980s, biologists began a forth synthesis under the name of evo-devo—evolutionary developmental biology. It is a work in progress.

This book is introductory, written for the general reader. It uses simple language, employs every-day examples, and avoids technical vocabulary. When biological terms are necessary, the text defines them, often more than once. A glossary of terms is also provided. Because the book covers only facts and concepts already familiar to biologists, it omits footnotes. However, it furnishes a bibliography. It also offers an extensive index. This book is designed to be short, user-friendly, and useful. It tells readers what they need to know about the growth of evolutionary biology. There is much more to learn than this book contains. For connections with religion, explore my Web site at www.theologyauthor.com.

1. GRASPING THE ESSENTIALS

Ernst Mayr is one of the architects of modern biology. He is also one of the greatest biologists of the twentieth century and one of the outstanding historians of biology. His magisterial work of 974 pages, *The Growth of Biological Thought,* was published in 1982. After an introduction on philosophy of biology, his history reaches back to the ancient philosophers and continues through the 1970s. By calling it the *growth* of biological thought, he emphasizes that scientific knowledge not only changes, but accumulates, as indeed it does.

Science accumulates knowledge

Science describes the physical world and discloses its history. The physical world is composed of matter and energy. Science limits its descriptions to matter and energy, their composition, interactions, and development. As it turns out, matter and energy are interchangeable by Albert Einstein's famous formula, $E=mc^2$. Energy is the foundation of the physical universe.

Science cannot describe the physical world exactly as it is, for science is ignorant of many details. Moreover, an exact description would be too cumbersome to use. Therefore, science builds simplified models of the physical world. In many ways, this is similar to creating a map. Maps are models. A city street map, for example, may show the main roads and the location of the central government buildings and places of historical interest. However, it cannot show all the details because these would clutter the map, making it impossible to use. Moreover, the map's purposes are restricted. Some of the features are irrelevant to the purpose of the map. Others may be unknown to the mapmakers. Thus, the map ignores the minor streets, the shops, apartment complexes, private homes, telephone poles, electric lines, and sewers. In comparison, $E=mc^2$ models matter and energy as interchangeable. *E* stands for energy, while *m* stands for matter. The *c* stands for the speed of light. The speed of light squared is a very large number. The formula says small packages of matter contain huge amounts of energy, as in atom bombs. It does not indicate the composition of matter, or why *c* is integral to the formula. This small model is very simple and very profound. It answers the questions Einstein posed about the relationship of matter and energy. However, it leaves unanswered many questions about matter and energy we might want to ask. The model is both simplified and restricted.

Scientific knowledge changes. However, it does not change randomly. Rather, it develops, accumulating facts, asking new questions, making fresh predictions, and altering and enlarging theories. Scientists may know certain facts A, B, and C and have a theory that ties them together. However, the theory is limited because it fails to cover other known facts, D and E, which may support a different theory. Through painstaking work, scientists develop a new theory that not only explains A, B, and C, but also explains D and E. This is a better theory because it is bigger. It encompasses more knowledge. At the same time, it may raise new questions, leading scientists to look for new information, to discover new facts. Later, these may be accommodated in the old theory, or they may require a theory that differs from the old one. Thus, scientific knowledge accumulates, and theories expand and change to encompass the new information.

For example, Charles Darwin, the founder of modern biology, knew several books that described evolution, one written by his own grandfather. The idea that species transformed into other species was popular, yet scientists rejected it because no one could provide a mechanism to explain how species change. During many years of observation and experimentation, Darwin discovered facts no one had known before, then added knowledge that others accumulated from geology, embryology, anatomy, and biogeography. (Biogeography is the study of the location on Earth of existing and fossil organisms.) Then he offered a mechanism that explained the facts. He called the mechanism that drove the transformation of species natural selection. Discovery of this mechanism for species change transformed the study of biology forever.

Darwin's work raised new questions. This is one sign of a major scientific theory. Major theories are fertile. They inspire new questions, open up new areas of research, generate new publications, and sometimes create new disciplines and new institutions. One of the major questions Darwin's work raised was about heredity. No one knew about the role genes play in heredity because no one knew about genes. Almost 50 years after Darwin's work, biologists discovered genes. At first, some biologists thought they raised problems for Darwin's theory of evolution by natural selection, but after several decades of theoretical work and new discoveries, biologists incorporated genes into the theory of evolution. The enlarged theory is called neo-Darwinism or the New Synthesis. It supports Darwin's mechanism of natural selection so well that for almost 75 years now, biologists have considered natural selection the main mechanism driving evolution, just as Darwin's original theory said.

A scientific theory, then, is not some wooly idea invented for the occasion. Rather it is a carefully worked out concept, based on evidence, used to integrate as much data as possible under one explanation. For example, before the discoveries of Isaac Newton, motions on Earth and motions in the heavens had separate explanations—fell under separate theories. Newton developed his theory of gravity to explain both motions using only one theory. His theory of gravity marked a huge advance in our understanding of the universe! Newton's theory suggested new research. It prompted the building of observatories and refinements in telescopes. Using his theory and improved telescopes, scientists discovered new planets that Newton's theory predicted. Thus, scientific theories perform important work. Sometimes, they seem so secure, so well-established, that scientists dub them laws. But in science, laws are theories and subject to change, like all theories. Einstein, for example, expanded and changed Newton's law of gravity to encompass fast-moving objects. Like all so-called laws in science, Newton's theory of gravity was open to change.

As science developed, it uncovered too many facts about the physical world for one person to grasp. The solution was to divide science into new disciplines, so that one person could learn all there was to learn in one discipline. Scientists began to specialize. The major divisions of the sciences dealing with the physical world, but excluding living organisms, are physics, the study of matter and energy as they are in themselves; cosmology, the study of the matter and energy as organized in the heavens; and geology, the study of matter and energy as organized on Earth in all that is lifeless. Biology is the study of matter and energy as organized in life forms.

In general, biology studies the development, anatomy, physiology, and behavior of all living organisms, including human beings. A new field in biology, evolutionary psychology, studies the evolutionary basis for human behavior. Biology also studies dead organisms and extinct species—fossils—and, so, it is a science that studies the history of organisms. Biology is a historical science. It studies change over time. That is to say, it studies evolution.

Evolution means change

Everyone is familiar with the development of individual organisms. Mammals and birds, for example, begin as small new arrivals, then grow, changing shape, increasing in weight, redistributing bulk, and adding hair or feathers until they reach adulthood. Development is predictable. Each organism seems to develop by a plan, a plan we now think of as under genetic control.

Evolution works differently. To compare it to individual development is a mistake. Individuals do not evolve. Only groups evolve. Evolution is a theory about populations, making it a statistical theory. Populations change over time as the individuals within them change in anatomy, physiology, and behavior. There is no plan. All change is local, caused by the environment in which the local population lives, including its ecological relationships with other organisms. The main mechanism driving population change is natural selection.

Natural selection is easy to understand because we see its components every day. We know that some characteristics of organisms are inherited, yet offspring do not exactly resemble their parents or siblings. Thus, in a litter of puppies whose parents are short-legged, generally quiet, and black and white, the puppies will also have these characteristics. Yet, some will be all black, some all white, and those of mixed hue will have black or white shading that differs from their parents and the other puppies. Some will have longer legs than others. Some will be more active than others. In the wild, they would be subject to disease and might become meals for other animals. Many, perhaps most, would die before reaching maturity. If the black ones were well camouflaged in their local environment, they would be more likely to survive than their white or spotted counterparts would. Eventually, as the white and spotted ones became food for predators, the local population of dogs might become all black.

The primary issue here is not survival, but survival to reproduce. Evolution is about inheritance down generations. If an organism survives but fails to reproduce, it will lack descendents in future generations and therefore will be irrelevant to evolution. Evolutionary changes are heredity changes in populations and, so, depend on reproduction.

Evolutionary changes in populations are well-known. During the nineteenth century industrial revolution in England, soot from burning coal covered the countryside. Populations of moths there had been multicolored, but hungry birds ate the lighter ones that stood out against the soot-darkened trees, leaving mostly black ones to reproduce. As a result, the population turned almost wholly black. Later, people cleaned up the environment, and the previously darkened trees returned to their natural, lighter color. Then the birds could best see and devour the black moths, so only the lighter ones lived to reproduce, and the population returned to its previous lighter color. Over several generations the population evolved twice in opposite directions, once to darker, then again to lighter. These are statistical changes within a population. The local environment shapes the direction of evolution. Features of a population can come and go. The direction of evolution displays no unalterable trend.

As the examples show, evolution by natural selection requires that organisms display only three general features, all related to reproduction. First, they must inherit some traits from their parents (heredity). Second, offspring must differ from their parents and each other in some traits (variation). And third, more offspring must be generated than survive to reproduce (multiplication). Heredity, variation, and multiplication characterize organisms everywhere in nature. In organisms that multiply profusely, as many egg-laying organisms do, each female may generate hundreds of offspring. Only a few may survive. In such cases, the pressure of natural selection is great, and evolution can occur quite rapidly.

Now we know genes carry the information that is inherited. The black puppies in the example have slightly different genetic structures than the white ones do. If all the white ones die before reproducing, the information in their genetic structures will die with them. It will disappear forever. If a change in genes occurs, genes are said to mutate. Organism carrying mutations that make them more successful in the local environment will survive to reproduce more successfully than the organisms without the mutation. Therefore, the mutation will spread throughout the population. If the mutation is lethal to the organism in which it appears, that organism with its mutation will quickly disappear from the population. If neutral to the organism's success, the mutation may persist down the generations, but spread slowly.

Thus, evolution might be considered a three-step process. First, mutations occur. These occur randomly, without respect to the needs of the organism. They may help it survive to reproduce or harm or kill it. Second, the mutational changes may begin a series of processes in the organism that result in observable traits. Because their mutational causes are random, these traits also occur randomly, without respect to the needs of the organism. Then natural selection acts on the organism. If the new traits help it survive to reproduce, it will thrive and create more of its kind. If not, it may decline or die without reproducing and its representatives in the population will decrease or cease. Thus, the population will evolve over time. It will become adapted to its environment.

The evolutionary changes discussed so far are small, and usually such small changes do not affect speciation. Speciation is the transformation of one species into another. Separate species, by definition, cannot breed with one another in nature down the generations. In order to form separate species, usually some members of a species must become isolated from the rest of the members. Often, only a few members become isolated and constitute a founder population. Because evolution will change both the founders and the parent species, over time the changes may become sufficient so that, if reunited in nature, the populations can no longer interbreed successful. They have speciated—become separate species. Underlying the changes are changes in genes.

Knowledge of genes led biologists to define evolution as changes in the occurrence of all the genes in a population over time. They refer to this briefly as changes in gene frequencies. However, it is important to remember that genes change organisms, and it is the whole organism that survives to reproduce or dies without offspring. Natural selection works directly on organisms. It selects genes only indirectly, through the organism. Modern biology has often placed too much emphasis on genes which, in the final analysis, are lifeless chemicals.

Genes are chemicals

Living organisms reproduce. They also metabolize, that is, they produce energy and assimilate new materials, casting off wastes. Chemicals do neither of these things. Chemicals are lifeless.

All chemicals are composed of atoms. Organisms use many chemicals, but the most important are hydrogen, the simplest atom, and carbon and oxygen, which are more complex. Seen another way, organisms are composed almost entirely of water. Water is two parts hydrogen and one part oxygen, the familiar H_2O. The other main chemical is carbon. The big bang, the event that startled our universe into being, produced all the hydrogen in the universe. The stars created all the oxygen and carbon in the universe. Therefore, all organisms have their origin in the heavens, human beings included. We are intimately united with the universe all the way back to the big bang, which occurred some 13.7 billion years ago.

Genes are merely chemicals. Their arrangement in living organisms is what makes them special. The arrangement of genes carries information passed on unchanged, generation after generation. This means genes are very stable. They are inactive. They have two functions: to preserve information and to provide templates for the production of proteins. Genes themselves do very little. Proteins act upon them to produce other proteins, using the inactive genes as templates.

Each cell in any organism's body has copies of all the genes the organism possesses. In theory, then, any cell could reproduce an entire organism. However, only some genes in each cell are used. The rest have no function. Heart cells use only the genes that will produce hearts to make hearts; skin cells use only the genes that will produce skin to make skin, and so on. Biologists say genes are turned on (or activated or expressed) if they are used. If they remain unused, they are turned off (or not activated, or not expressed). Through the production of proteins, some genes regulate other genes, turning some on, others off.

Known as regulatory genes, these genes are very important, for they not only regulate the genes they turn on and off, but those genes produce proteins that influence other genes in an extended cascade of activity.

Environments are important, too, for they also turn genes on and off. For example, arrow leaf plants grow both in water and on dry land. If they develop in water, their leaves are narrow and flexible. On land their leaves are broad and rigid. This is because the watery environment activates some genes while silencing others. On dry land, the dry environment activates different genes. In another example, some caterpillars that eat flowers resemble the flowers they eat. If they eat twigs, they resemble twigs. Here, the environment turns genes off and on that are important to camouflage the caterpillars to look like the plants they live on. Genes in people also respond to their environment. If children hear no language until they are adolescents, they fail to develop normal language, for the genes that a linguistically charged environment would turn on are never activated. If they never walk, or if their feet are bound, an ancient Chinese practice, the bones of their feet fail to develop for efficient walking. Thus, it is incorrect to think that genes alone determine the outcome of development in living things. Genes and environments everywhere act in concert to produce mature organisms.

Genes also act in concert with other genes to create organisms. They work together so the organism can function. They collaborate to fight dangerous germs that invade the organism. Remarkably, it has become popular to talk of genes as selfish. What a funny idea! Genes have no selves. They are lifeless chemicals. Moreover, they cooperate with other genes. If they ceased to do so, the organism would fall apart. When a few genes do cease to cooperate with others, cancer ravages the organism. Selfish genes are diseased genes. All healthy genes are selfless, working as a team to develop organisms and keep them functioning. However, it is best not to apply such misleading terms as selfish and selfless to genes. These terms apply to people. Genes are not human. They are chemicals. They have no motives. They do almost nothing. Proteins are the active agents in organisms. However, they, too, lack life and motives.

Summary

Science describes the physical world, building simplified models and explanatory theories that help us understand how the physical world developed and how it functions now. Science changes as scientists discover new facts and raise new questions. Changes in knowledge of biology inspired Darwin to understand how populations of organisms transform over time through evolution. The New Synthesis merged genes into the theory of evolution which, in turn, enhanced scientific understanding of Darwin's original mechanism of species change, natural selection. Recently, biologists discovered that some genes regulate other genes. How this happens remains unclear. Thus, as science progresses, it raises new questions and makes new predictions that, as they are addressed, create other questions and predictions. The advancement of science may result in the creation of new publications, instruments and institutions. Science does not merely change. It accumulates knowledge. Human beings guide the accumulation.

Evolution, too, means change. However, in evolution, there is no guide. Evolution is local, changing populations wherever three features occur: traits of organisms are inherited, offspring vary from their parents and each other, and more organisms are generated than survive to reproduce. With the discovery of genes, biologists saw evolution as a change in gene frequencies in populations.

Genes are lifeless chemicals that carry information and serve as templates for proteins. Environments influence genes, turning some on, others off. Genes themselves also produce proteins that regulate other genes, turning some on, others off. Genes cooperate for the development and maintenance of organisms. They, and the organisms they produce, are composed of elements made in the big bang and in the stars. Life is intimately tied to the rest of the universe, for its components were made there. In living forms, these components are structured so organisms metabolize and reproduce, performing the activities that characterize life.

Because almost everything in biology has exceptions, nearly everything in this chapter has exceptions.

2. UNDERSTANDING EVOLUTION

Charles Darwin published *On the Origin of Species by Means of Natural Selection* in 1859. It ran 490 pages. He followed it in 1871 with an 827-page volume, *The Descent of Man and Selection in Relation to Sex.* This chapter views evolution much as Darwin did, although for ease of explanation, it uses modern concepts unknown to Darwin, like genes, viruses, and the components of cells.

Evolution means change

Evolution is simply another word for change. The idea that species change goes back to the ancient Greeks, and Europeans were familiar with it before Darwin. Species evolve as populations, not individuals. And, unlike the development of individual organisms, evolution in populations has no plan, no guidance such as genes provide for individual development. Changes are random with respect to the good of the organism or population. The mechanism of natural selection saves those best adapted, so they thrive and multiply.

In most cases, populations evolve slowly. Some species stay stable for millions of years. Evolution is slow and gradual because it often depends on small changes that must spread through a population by breeding. Therefore, all other things being equal, populations with females that bear many young, and bear them frequently, evolve faster than populations in which breeding is slow. Populations of flies can evolve rapidly. Populations of elephants cannot. This is one reason biologists like to perform evolutionary experiments in the laboratory on fruit flies and, for mammals, on mice. They breed rapidly and therefore change rapidly. Of course, fruit flies and mice are also cheaper to house, feed, and rid of droppings than are elephants!

Creatures of our physical size and inability directly to observe microscopic organisms, scanning the broad sweep of evolution over four billion years, tend to see a direction. Life begins and proceeds for about two billion years with very simple, single-celled organisms that lack a nucleus. They are the bacteria.

Then life becomes more complex as some of the single cells acquire a nucleus and internal structures. In technical terms, prokaryotes, cells lacking a nucleus, become eukaryotes, cells with a nucleus. Later, some eukaryotes join together to form multicellular organisms, organisms we can finally observe directly. These, in their turn, evolve into plants and animals. Some of these later evolved into the organisms familiar to everyone—grasses and oaks; lions, wolves, and zebras; ants, bees, squirrels, and mice, as well as organisms whose evolution we ourselves shaped, like corn and tomatoes, cows, sheep, horses, and dogs. From this perspective, human beings appear to constitute the climax of evolution on Earth. We are the most complex, the most intelligent, and the most cultural of all Earth's creatures. Yet, the bulk of organisms on Earth today are still, by far, bacteria. Is, then, the direction of evolution toward us, or do bacteria continue to be its favored creatures? Furthermore, Earth will exist for at least another four billion years. What will evolve during that vast future? Will we, like most species, go extinct? Will species evolve superior to us in complexity, intelligence, and cultural ability? Given the possibility of nuclear war, the story of evolution on Earth may proceed from the simple cells of bacteria back to the simple cells of bacteria. Then, multicellular organisms that thrived and diversified will constitute a mere interlude in the triumph of non-nucleated cells.

The consensus among biologists today is that evolution lacks an over-all direction. It is local. It occurs in populations of limited size, in particular places, and at specific times. It depends on many things, changing environments being one of them. In some places and at some times, environments change rapidly. At other times and places, they remain stable for millions of years. In stable environments, evolution tends to stagnate. In changing environments, complexity, sensory capacities, and intelligence may increase, or they may decrease. The most obvious examples of decrease occur in organisms that originally lived in sun-lit environments, but moved into caves. Over time, all these organisms tend to lose their eyesight—and eventually their eyes. Nothing in their perpetually dark environment keeps sight functioning. As a result, vision wastes away. Other animals, like dolphins and whales, lost their legs as they moved from land, where legs are useful, into deep water, where they are not. Evolution can reduce complexity as well as enhance it. In all these cases, the mechanism of natural selection is acting. It adapts organisms to their environments, whether by adding traits or removing them. Darwin's favorite phrase for these changes was "descent with modification." As offspring descended from parents, as generation descended from generation, natural selection retained or eliminated various modified traits.

Although evolution is local and unpredictable, it works logically. The logic is *if-then* logic. *If* three necessary features are present in a population, *then* the population will change over time, all other things being equal. If some traits are inherited, if offspring vary from their parents and each other, if more offspring are generated than survive to reproduce, then a population will change over time. It will evolve. Evolution in this sense is predictable.

Consider a population of deer. Among them, individuals differ. Some run faster than others, for example. Wolves that eat deer inhabit the same territory. The faster deer can escape the wolves, while the slower ones become meals. As this process proceeds, the wolves cull out the slower deer. They fail to survive to reproduce. Now the population of deer consists only of the faster ones. The population has changed. It has evolved. The mechanism of natural selection has adapted the surviving population to the presence of speedy predators.

Of course, factors other than wolves act on the population. The deer need food and water. They harbor parasites. They suffer from numerous diseases. They require a somewhat temperate climate. And so on. Because so many factors interact to affect evolution, its direction is difficult or impossible to predict. Changes in any of these factors tend to drive the evolution of a population by natural selection, sometimes to extinction. However, changes in an interbreeding population do not usually lead to speciation. To form a new species, some members of a population typically must become isolated from the other members.

Speciation involves isolation

It was no accident that Darwin began to think in terms of species transformation when he explored the Galapagos Islands. These islands lie off the west coast of South America, right at the equator. They contained species found nowhere else on Earth. Darwin found no animals there whose eggs and larva immediately die in salt water. He found no mammals except bats, flying mammals. He found many species of birds, but mostly marine species. The many plants he saw had seeds easily transported by birds. He noticed the species on the islands resembled species on the mainland. And the island species resembled each other. Apparently, animals that can fly and the seeds they may carry with them managed to move from the mainland to the islands. In contrast, those salt water destroyed or whose swimming was marginal, as with most mammals, remained on the mainland.

By definition, species are populations whose members interbreed with each other, but fail to interbreed with other populations in nature. *In nature* is important, for people can manipulate some species to interbreed by overcoming or bypassing factors that isolate species in nature. Flowering plants, for example, may bloom at different times in nature, and so never exchange pollen. Yet people can collect pollen from the early bloomers and pollinate the later bloomers with it, forcing interbreeding. Such populations are still considered separate species.

If members of a population remain together, they can interbreed with each other and, so, continue to distribute genes and genetic changes throughout the population. They will remain similar enough to continue to interbreed.

However, perhaps a small number of the members become isolated—say a storm blows a few birds from the mainland of South America to one of the Galapagos Islands, and they have no way to return to the mainland. Two factors in particular work to change this founder population rapidly, so it differs from the mainland population.

First, the founder population is statistically different from the mainland population. By definition, the mainland population is represented by all its members. The founder population is small and, therefore, unrepresentative. It lacks many genes found in the mainland population. Therefore, it also differs from it in some traits. The founder population may have shorter wings and be weaker fliers—one reason the storm blew them out to sea. They are likely to harbor fewer species of parasite than the population taken as a whole, for there are fewer birds. Their coloration may not be the norm. Perhaps these birds are greener and entirely lack a yellow wing-tip that some of the mainland population display. Stated mathematically, the differences constitute a sampling error.

Second, they arrive in a new habitat. On the islands, they may lack predators, but also familiar foods. The island climate may slow parasite growth or encourage it. Water sources may be more abundant or less. Nesting material will be different, whether better or worse. If they are the first birds to inhabit the island, they will have numerous niches to occupy, whereas on the mainland, competitor species occupied many good nesting and feeding sites. With the pressure of the new environment, natural selection begins to operate on the population. It will determine whether the birds die or thrive.

If they lack edible food or drinkable water, they will die. Indeed, death is likely, given the strangeness of the environment. On the other hand, they may arrive in bird paradise and reproduce rapidly, finding abundant food and water, nesting materials, and nesting sites. Without predators on the ground, they may even become flightless, as many actual island species did, diverting the extra energy to enhanced reproduction. As they evolve, they will become increasing dissimilar from their mainland counterparts, unable to breed with them if returned to the mainland. They will have become a separate species.

Yet, speciation rarely requires drastic changes. It may occur simply because members of one population cease to recognize members of another as mates, whether because of coloration, size, or other matters of appearance. Courtship behavior may drift apart in movement or song. Sex organs may be incompatible. If compatible, the offspring may be aborted or be born dead or may be sterile. Breeding seasons may differ. Even small changes can make interbreeding unlikely or impossible. As Darwin stressed repeatedly, natural selection can see the smallest trait. As biologists now understand, speciation is an evolutionary accident, the byproduct of mating barriers and physical isolation.

Islands are unnecessary to provide for physical isolation. Mountains also isolate members of a population from one another, as do rivers, rock slides, lava flows, highways, food choices, and breeding sites. This simple model for speciation resulted in much research, which is still occurring today. The model has proved its worth!

Darwin provided another model, one that does not depend on the isolation of populations. Instead, it depends on choice. It is sexual selection. Sexual selection is only now being investigated carefully. It is the preference of one sex, usually the female, for certain traits of the opposite sex. The female may reject a potential mate merely because she finds him unattractive, although he is otherwise fit for reproduction. However, recent studies suggest that males that females find attractive may appear attractive just because they are superior reproducers. She does not know this. But those females who chose the traits marking poor reproducers had fewer and sicklier offspring that failed to live to reproduce down the generations. Those attracted to traits marking good reproducers produced attractive and powerful offspring that survived to reproduce down the generations. The genes that helped females be attracted to good reproducers survived in their female offspring, who were also attracted to good reproducers down the generations.

Understanding sexual selection fully requires some background. Males, by definition, produce sperm, whereas females, by definition, produce eggs. Sperm are small and created in abundance. Males use little energy to make them. In contrast, eggs are few and large. They require considerable energy to create. Moreover, in animals that carry their young in pouches or wombs, the female carries them in almost all cases. (Seahorses are an exception.) Often, too, the young depend upon their mother for weeks, months, or years after release from the pouch or womb while the male roams free in search of other mates. In brief, females expend lots of energy and resources to generate offspring and raise them to reproductive age. Males expend little, less than the female even in cases where both work together to raise their mutual offspring.

Because reproduction costs the male little, evolutionary theory predicts that he can afford to be causal about his mating. Males mate with anything they perceive as receptive females, and even mistakes of perception usually cost little. On the other hand, as evolutionary theory predicts the female must be careful of her mates. An inferior mate may mean she uses all her reproductive energy to generate offspring that never attain reproductive age. In evolutionary terms, she is a loser. Copies of her genes disappear from the population. She leaves nothing to posterity. And having copies of genes passed down the generations is important. Otherwise, all life would go extinct.

Therefore, predictably, in sexual selection females do the selecting. Their mate preferences affected the outstanding features of male birds: their brilliant coloration, including the peacock's tail, and their elaborate songs. (The songs may also play a role in male-male competition.) Without mates, the bland and silent males' traits disappeared from the population.

Sexual selection is one cause of speciation. Even some females in a population who select mates of one color, whereas others select mates of another, can initiate speciation. Such is the case today with Cichlid fish in Lake Victoria in Africa. Some females choose males that appear redder. Others select ones that appear bluer. These actions affect the gene pools of each population. A gene pool is the collective genes of an interbreeding population. Here, the gene pools of the two populations have begun to separate and differ. Interbreeding is rare. In this lake, biologists are watching speciation happen.

To speak of female choice may sound as if the female exercises thought and foresight and knows about genes. Of course not! Plants and most animals lack both thought and motives, and even most human beings are ignorant of genetics. Evolution occurs because of the consequences of dissimilar rates of survival and reproduction. If, for whatever cause, females mate only with the brighter-colored males, those males will reproduce successfully. The dull males will fail to find mates and their genes for dull color will disappear from the population. Over time, if only the brightest-colored males reproduce successfully, the population of males will grow more and more brightly colored.

Sexual selection once more provides a simple model of what happens in nature. During the past quarter century, it has been the focus of considerable research. It is a within-species example of one of the strongest drivers of evolution, coevolution. Sexual selection differs from coevolution only because biologists define coevolution as occurring between species rather than within them.

Coevolution, then, causes changes in two or more species whose interactions influence each other. The most familiar examples of coevolution are predator-prey relationships. The case of wolves causing a deer population to become faster over time involves two species, not merely one. The other is the wolf population. If the deer population becomes fleet enough, and the wolf population fails to evolve, eventually the wolf population will starve, all other things being equal.

More likely, as the deer population becomes fleeter, only the slower wolves starve. The faster ones eat well and reproduce successfully, multiplying fast wolves. Thus over time the wolf population will track the deer population: both become faster. Thus, each population is acting as a selection mechanism on the other. The wolves select the slower deer by eating them. The deer select the slower wolves by leaving them behind to starve.

Of course, the example is over-simplified. As the deer population changes due to pressure from the hungry wolves, speed among the wolves might be insufficient to maintain the wolf population. It might die out, or the wolves might evolve methods of cooperative hunting, with ways to surround a small number of deer and attack them as a pack, then share the meal in peace. This complex example requires many changes in the wolf population. First, as individuals ceased to be effective hunters, the wolves must hunt in groups. Second, they must coordinate their hunting to be successful. Third, they must restrain their individual tendencies to grab all the food for themselves. Because they do not plan their evolution, this again is a case of dissimilar rates of survival and reproduction. Those that tended to hunt together ate well and multiplied. The loners starved. The remaining wolves improved their hunting in packs as generations succeeded one another. Because many species of animals evolved to hunt in packs, the necessary tactics evolved many times over the eons. This is another example of descent with modification of traits.

Biologists think of the coevolution of predator and prey as a kind of arms race. As members of one species acquire a trait that helps them elude or capture members of another, the other species must acquire a trait that helps them capture their prey or elude predators. Other modes of arms races occur, too, of course. An arms race can enhance almost any trait, including chemical traits, as in poisonous snakes. Plants, too engage in chemical arms races, producing substances toxic to the insects that devour them. Successful insects evolve to withstand the poison. Again, none shows foresight or makes plans. The resistant organisms merely survive and multiply. Moreover, when trait-acquisition depends on the occurrence of a new mutation, the mutation may or may not appear. Mutations have neither motives nor purposes. They do not occur for the goal of helping an organism or population or species. They merely occur. Some do help. Others are neutral, neither helping nor harming. Others cause harm. In all cases, their occurrence is random with respect to the needs of the organism or population or species.

Another coevolutionary mechanism is parasitism. Parasites are organisms that live off other organisms to the hosts' harm. Viruses are parasites of a special kind. Once they were living cells, but now they are simple molecules, not organisms. They constitute another example of evolution acting to reduce complexity. Viruses are not even alive. This is descent with modification of traits with a vengeance! Viruses can metabolize and reproduce only by seizing the host's cellular machinery. In doing so, they damage the host.

Some parasites and viruses are quite lethal. However, killing the host is a poor evolutionary strategy, for the parasites' means of survival dies with the host. Unless the parasite can find another host quickly or survive without its host, it too will die. This means that extremely lethal parasites often go extinct, to be survived by their less lethal relatives. These less lethal parasites survive to reproduce on a host that may suffer damage, but lives on. Thus, on the whole, evolution by natural selection works to make any given species of parasite less lethal over time. It adapts the parasite to the host.

For organisms with extremely short lives, parasites pose little problem. The host dies, and so does the parasite. If the parasite dies before it can reproduce, neither it nor its potential offspring can infect another host. However, organisms with longer lives provide good hosts for parasites. Many biologists think sex evolved at least partly to decrease the problem of parasites. Why? Because sex mixes genes. The mixture means offspring differ from their parents in many respects, and therefore the parasites able to thrive on the parent may find the offspring a less supportive host. It is interesting to consider that our ancient ancestors' coevolution with parasites may be one cause of our being male and female!

The most stable mechanism for coevolution is mutualism, which is also known as symbiosis. In mutualism/symbiosis, different species live together for their mutual benefit. They are mutually well-adapted. Sometimes, they live together so closely, they unite to become one organism and form a new species. Because mutualism/symbiosis is more stable than parasitism is, many parasitic relationships evolved into mutual or symbiotic relationships between species, to the mutual benefit of both. People with only a little knowledge of biology can identify some symbiotic relationships. Insects, bats, and birds pollinate flowering plants. Fungi and algae together form lichens. Leaf-cutter ants farm fungi. Human beings domesticate plants and animals.

Many mutualistic relationships affected the evolution of life on Earth. The earliest mutualism occurred among bacteria, each a single cell. They formed a common gene pool, similar to that formed by a single species today. In a single species, the organisms in the species exchange genes. These interchangeable genes constitute the gene pool. Because bacteria exchange genes with one another, their genes also constitute one gene pool. Bacteria are mutualists, exchanging genes, benefiting from their common lives with other bacteria. Biologists consider bacteria to belong to the same species if they share 70 percent of their genes. If they share fewer, biologists consider them separate species.

Perhaps the most amazing example of mutualism was first described by Lynn Margulis in the 1960s and 1970s. It happened with the evolution of the eukaryotic cell. Bacteria are prokaryotes. Their cells are simple, lacking a nucleus or, indeed, much structure at all. Eukaryotes are complex cells, with a nucleus, an internal structure, and other "little organisms" inside them, known as organelles. Evidence now confirms that the nucleus and organelles of eukaryotes were once separate cells of single bacteria. Almost all human cells are eukaryotic. If we wish, we can think of ourselves as composed of bacteria that evolved to live together for their mutual benefit—and ours.

The most familiar examples of mutualism are multicellular organisms. All the organisms we can see are multicellular. In multicellular organisms, cells specialize, some forming the skin, some constituting the scales or hair or feathers, some making livers, some living as neurons in the brain. In multicellular organisms, all the cells must live in a mutually cooperative relationship for their own benefit and that of the organism as a whole. If they cease to have mutually beneficial relationships, the organism falls apart. It ceases to exist. And with its death, all its cells die.

Plants evolved by mutualism. Some 450 million years ago, algae and fungi fused, creating plants. Plants that evolved since then have thrived because of that symbiotic union. Today, approximately 90 percent of land plants form other symbiotic relationships with fungi. They supply fungi with sugars, while fungi bring them nitrogen and phosphates, and sometimes antibiotic and drought protection, too.

Insects that eat plants form symbiotic relationships with bacteria to help them digest the cellulose the plants provide. Mammals that eat plants also require bacterial symbionts. (Symbiont is the technical term for organisms living in symbiosis, especially the smaller one.) Cows, of course, are famous for the bacterial symbionts that aid their digestion. But sheep and goats and deer and antelope and people also house bacterial symbionts that assist digestion. Hundreds of bacterial species occupy our digestive systems. When we take antibiotics to kill harmful bacteria that sicken us, we often upset the balance of helpful bacteria in our digestive systems, adding to our misery. Some people now eat probiotics, beneficial bacteria, to restore the balance. However, we have insufficient knowledge at present to know how all these symbionts work for us, and how to keep them healthy, and therefore how to keep us healthy. If we wish, we can think of ourselves as a giant city, with many species of bacterial workers that we keep alive by providing them with shelter and food, while they keep us alive by helping us digest our food and defending us from their harmful relatives.

Aside from this unplanned mutualism with bacteria, we formed deliberate symbiotic relationships with other organisms by domesticating them. We bred them over millennia to serve our purposes. Domesticated rice and wheat constitute staple crops for billions of people. Some domesticated animals, like horses, can survive in domestication or thrive in the wild. Others, like cows and chickens, depend on us for their survival—and we on them. The simple scientific model of coevolution has produced rich results in our understanding of evolution.

Summary

Evolution refers to changes in populations. Because changes occur down the generations by descent with modification of traits retained or discarded by natural selection, usually evolution proceeds slowly. It occurs most slowly in organisms having few offspring years apart and happens most rapidly in organisms that have hundreds of offspring in a matter of days or months. Although no one can predict its details, the process of evolution is logical, requiring only three necessary features: heredity, variation, and multiplication. This very simple model of evolution has revolutionized biology. It is still the prevailing model. Indeed, after a century and a half of research on other models, it is the only one to survive.

Speciation is a special kind of evolution. It creates new species. Usually, it requires that a few members of a species become isolated from the others. Isolating factors include islands, mountains, lava flows, and highways, food choice and mate choice. Because the founder population is statistically different from the main population, it is likely to evolve in a different direction. Because its habitat differs from the main population, the pressures of natural selection on it may be great, and it may evolve quite rapidly into another species. Speciation through isolation and natural selection accounts for many of the species we see around us today.

Coevolution constitutes one of the main drivers of evolution. In coevolution, one species affects the evolution of another. Coevolution may occur through predator-prey relationships, parasitism, or mutualism/symbiosis. Predator-prey relationships produce a kind of arms race, pushing each species to evolve in order to maintain its balance in relationship to another species. Here, the two species provide reciprocal selection forces. Parasitism, in which an organism harms or kills its host, tends to be unstable and often evolves into mutualism/symbiosis. Mutualism provides an important source of speciation in the history of evolution. By mutualism, prokaryotic cells joined together to become eukaryotic cells. By mutualism, lichens formed. By mutualism, plants evolved. Today, most land plants depend on further mutualism/symbiosis with fungi. We, too, depend on symbiosis. We domesticated plants and animals for food. Bacteria in our digestive tracts help us digest our meals. Furthermore, bacteria live throughout our bodies, helping us in innumerable ways. They out-number the cells in our bodies by ten to one. These multitudes of bacteria and our domesticated stocks enable us to survive in mutually beneficial relationships. Natural selection has adapted us to aid one another.

Because almost everything in biology has exceptions, nearly everything in this chapter has exceptions.

3. CALCULATING GENETICS

In 1942, Julian Huxley published *Evolution: The Modern Synthesis,* a work of some 645 pages. It united Darwin's evolutionary mechanism of natural selection with analysis of evolutionary trends and processes and their relationship to genetics and classification. Huxley's book provided the common name for the growth of Darwin's work during most of the twentieth century, the Modern/New Synthesis, also called neo-Darwinism. Most biologists accepted the New Synthesis by 1947.

Genes supply the mechanism for heredity

Darwin's original theory required three features: variation, multiplication, and heredity. But although plain observation showed organisms inherited some traits in each generation, no one knew how heredity came about. No convincing mechanism existed. Several biologists speculated about the mechanisms. Darwin called them gemmules. Others named them determinants, pangenes, or biophors. Each biologist based his idea on some experimental evidence, but none of the proposed mechanisms seemed to provide an adequate explanation of heredity.

Although clergy denounced Darwin's theory for religious reasons, and physicists criticized it on cosmological grounds, heredity was the problem that kept Darwin awake at night and forced him to reconsider his account of evolution. He always considered natural selection the main mechanism for evolution. However, he included a popular heredity mechanism, "use and disuse," in his theory. Today his "use and disuse" is known as the inheritance of acquired characteristics. According to this view, using a trait enhanced it in the user, and subsequent generations inherited the improvement. The blacksmith hammered iron. His arms grew muscular. The giraffe reached for leaves high in the tree. Its neck grew longer. If offspring inherited these acquired characteristics, then the blacksmith's children would be muscular, the necks of the next generation of giraffes longer. Disuse reduced or eliminated the trait in question. Under the pressure of criticism regarding his ideas about inheritance, Darwin increasingly emphasized this idea as new editions of *On the Origin of Species* appeared.

The other concept of heredity common at the time was blending inheritance. It seemed obvious. Two parents, one five feet tall, the other six feet, often produced offspring that ranged between the two heights. Their heights appeared to blend in their children. Other traits appeared to blend in offspring as well. However, if inherited traits blended, then traits would merge in each generation rather than remaining stable. And if traits merged, Darwin's theory faced difficulties, for evolutionary changes depend on the stability of traits into the next generation. If they blended, evolution would cease over time.

Shortly after the publication of the *Origin* in 1859, an Austrian monk named Gregor Mendel published a paper on inheritance in an obscure scientific journal. Mendel performed careful experiments on peas and kept meticulous records. Had Darwin known of Mendel's work, he would have slept soundly. However, Mendel's paper lay unappreciated for 34 years. When biologists working on the problem of inheritance discovered Mendel's paper in 1900, they found it contained information that transformed their understanding of inheritance and made it predictable.

Mendel's model was simple. Parents differ in their genetic constitutions. Each contributes an equal number of genes to their mutual offspring. Genes never blend. Rather, they transmit separately as discrete units down the generations. Furthermore, one and only one gene underlies each trait.

Mendel's simple model solved Darwin's problem. Inheritance is particulate, not blending. Mendel's work also suggested that acquired characteristics are never inherited. Separate experiments performed around 1900 proved this correct. The most noted experiment severed tails of mice for generations (ouch!). Yet, each subsequent generation appeared with tail intact. Biologists developed a vocabulary to divide observable traits from the unseen genes that underlay inheritance. The bodily, observable, traits of the organism they called the phenotype. The underlying genes considered together became the genotype.

Mendel's simple theory accounted for the experimental and observational evidence. Furthermore, it allowed mathematically inclined biologists to develop statistical formulas for inheritance within populations (population genetics) and down the generations (transmission genetics) that allowed them to predict certain aspects of genetic change. By 1947, biologists had combined these two forms of mathematical genetics with natural selection to explain adaptation, speciation, and fossil history. This combination forms the New Synthesis, the fundamental theory of modern biology. Biologists of the Synthesis modified Darwin's descriptive phrase for his theory, "descent with modification" of traits. It now read, descent with modification of gene frequencies. The modified theory constitutes a growth of Darwin's work. Its main contributors in addition to Julian Huxley are Theodosius Dobzhansky with *Genetics and the Origin of Species* (1937), the foundational text; Ernst Mayr, *Systematics and the Origin of Species* (1942), on the definition and classification of species; and George Gaylord Simpson, *Tempo and Mode in Evolution* (1944), on fossils.

Nonetheless, for several years some biologists questioned the physical reality of genes. Certainly, mathematical formulas for gene transmission were useful in making predictions. However, no one had seen genes, and no one knew what they were. Perhaps they were merely mathematical phantoms. Were there real, physical entities to which the term *gene* referred? The simple answer turned out to be yes.

Genes possess physical reality

No one knew about the chemistry of genes when Huxley published his work. Their structure remained unimagined. Nonetheless, experiments disclosed some facts about the material of inheritance that corresponded with Mendel's discoveries. Biologists had seen chromosomes, which they named for the ease of coloring them in live organisms. Chromosomes form pairs, one from the mother, the other from the father, as Mendel's work suggested. Mendel's peas had 7 sets of chromosomes, and we have 23. Biologists discovered that the chromosomes, one from the father, the other from the mother, engage in crossing over. That is, each chromosome in the pair could break, and when they reattached, rather than the father's reattaching to the broken end of his chromosome, it might attach to the mother's. This plus the independent movement of genes constitutes recombination. Recombination forms new gene combinations in offspring that do not occur in their parents. It scrambles genes, which we now know lie along the chromosomes, increasing the potential variety of traits in the organism. This increased variety provides material for natural selection.

The other way to enhance variety is through mutation. The New Synthesis emphasized mutations in single genes responsible for changes in the phenotype of the organism. Mutations provide traits new to the organism. These discoveries confirmed Darwin's prediction that evolutionary changes never occur due to the needs of the organism. Rather, they occur randomly with respect to its welfare. They may harm it, help it, or have no effect. Genes, then, according to the New Synthesis, are the units involved in recombination and mutation at the level of the genotype, whose chemical nature and structure were still unknown. They are also the units providing function at the observable level of the phenotype. However, evidence began to accumulate that genes cannot be all three things—the units of recombination, mutation, and function—in any simple way. Ambiguity lurks at the heart of the Synthesis.

Genes prove ambiguous

Although biologists still searched for the material nature of the gene, the genetics that worked was mathematical. Mathematical genetics depended on statistics and formulas for the mutation and movement of genes. As the founders of mathematical genetics knew, the mathematics failed to be firmly grounded in physical reality. One of the founders placed biology third in a list of importance, the first two items on the list being statistics and statistics! They followed Mendel who observed traits and inferred that unseen genes underlay the observed trait. Thus, he speculated that there were genes for wrinkled peas and other genes for smooth peas because he saw wrinkled and smooth peas. Because these traits reappeared intact down the generations, he assumed an underlying cause. In modern terms, he observed the phenotype and inferred the genotype. He never observed genes. Neither did the founders of mathematical genetics.

Experiments with plants proved that one gene, one trait, commonly known as "genes for" a given trait failed to explain experimental results. Plants provide wonderful experimental organisms. They stay in one place. Some generate rapidly and plentifully. They are easily cloned, producing numerous plants with the same genotype. Experimental biologists planted clones in different environments and observed that different traits appeared in the phenotypes, depending on the environment. Because every plant possessed copies of the same genes, the observed differences could not be related back to differences in genes. The same genes seemed able to produce a variety of traits. The arrow leaf plant became a textbook example. It grew in water and on land, and the dissimilar environments produced dissimilar phenotypes. Here was proof that phenotypes do not arise directly from genotypes in any simple way.

Biologists also learned to experiment with mutations. They discovered they could deliberately mutate genes through the use of chemicals and/or radiation. When they did so, the mutations produced complex results. Mutations in one gene sometimes affected more than one trait. Mutations of several genes sometimes affected only one trait—or none. These experiments also demonstrated that phenotypes do not arise directly from genotypes.

Twenty-first century biology highlighted the disconnection between genotype and phenotype. In 2001, the Human Genome Project showed that people have some 22,500 genes. Biologists already knew that bacteria have about four thousand. Yet, we are far more than six times as complex as bacteria. One gene, one trait, or "genes for" a given trait, cannot be true. Something else is going on. For this reason, the expression "genes for" will retain the quotation marks that signal its literal falsity.

Experiments disclosed both recombination and mutation. Both produced chemical changes in real, physical material in the nucleus of cells, but biologists knew little about the nucleus or the chemical nature of genes. They knew even less about development in single organisms. Genes underlay development—real, chemical genes. But their action was veiled in complexity. Therefore, throughout the twentieth century from Huxley's publication in 1942, the New Synthesis depended on mathematical genetics. Mathematical genetics supported natural selection as the main driver of evolution. It banished the problems with inheritance that haunted nineteenth century evolutionary theory. It made accurate predictions. It applied to problems of speciation, adaptation, and the biological history written in the fossils. The mathematical genes of the New Synthesis proved an enormous success. They constituted an important new growth to Darwin's theory of evolution and increased its ability to predict. Because the theory of evolution is a theory about changes in populations, it is a statistical theory. Mathematical genetics supplied the statistics and made successful predictions. Under the union of genetic inheritance and natural selection that the New Synthesis provided, the science of biology flourished.

Summary

Charles Darwin's theory of evolution by natural selection held deep insights. It revolutionized biology. It provided answers to age-old questions and drew relationships among fields of science that had seemed unrelated, from physics and geology to biogeography and embryology. It predicted by simple logic that evolution occurs. It raised new issues that produced large and successful research programs. These are all signs of a major scientific theory. Indeed, evolution is the only theory that encompasses all of biology. Without it, little in biology makes sense.

Nonetheless, the understanding of organisms remained undeveloped in Darwin's day. Of the three necessary features for organisms to evolve, one proved particularly troubling for his theory. This was inheritance. Nineteenth century speculations about inheritance challenged Darwin's theory. However, they proved wrong. At the turn of the century, Mendel's experiments, careful records, and published paper successfully addressed Darwin's concerns. Mendel's experiments and others on inheritance showed inherited traits appear and reappear intact down the generations with statistical and predictable regularity. Mendel and his fellow experimental biologists inferred that the underlying causes were discrete genes. Mathematically inclined biologists developed formulas that showed how genes could spread through populations and emerge unchanged down the generations. Their findings explained adaptation, speciation, and the evolutionary history revealed in the fossils. This synthesis of Darwin's theory with genetics confirmed the centrality of Darwin's main mechanism for evolution, natural selection.

Moreover, experiments began to demonstrate that genes were real, chemical entities, not mere mathematical phantoms.

However, these chemical genes failed to act exactly as the mathematical formulas of mathematical genetics predicted. Clones of plants grown in different environments produced different phenotypes. A single mutation refused to result always in changes in a single trait. Sometimes, one gene affected several traits. At other times, several genes affected only one trait. One gene, one trait, the foundation of mathematical genetics, could not be right. Early during the New Synthesis, biologists realized that "genes for" given traits did not exist in any simple sense.

Nonetheless, the New Synthesis proved an enormous success. It is the standard model in modern biology, taught everywhere. It predicted evolutionary outcomes, inspired new research, opened new publications, and altered institutions. Indeed, biology is probably the most fertile science of the last half of the twentieth century. The New Synthesis transformed Darwin's theory of descent with modification of traits into descent with modification of gene frequencies.

One of the flourishing triumphs of the New Synthesis is the explanation of the social behaviors of thousands of species of animals, based on discoveries in mathematical genetics. This is the subject of the following chapter.

Because almost everything in biology has exceptions, nearly everything in this chapter has exceptions.

4: EXPLAINING SOCIAL BEHAVIOR

Recognition of the biological foundations of social behavior arrived in 1975 when E. O. Wilson published his 697 page book, *Sociobiology: The New Synthesis.* As its subtitle indicates, he considered it a substantial contribution to the New Synthesis. It was. It demonstrated the power of genetic relatedness in promoting social behavior in hundreds of social species, including human beings. Its simple predictions inspired new research projects. The research proved fruitful, leading to new publications and the rise of journals, organizations, and institutions devoted to sociobiology.

Social behavior promotes multiplication of kin

Charles Darwin's theory of evolution by natural selection required that organisms possess three features: heredity, multiplication, and variation. The New Synthesis corrected nineteenth century speculations about heredity through the development of mathematical genetics and increasing knowledge of the chemical gene. Early in the century, chemists discovered DNA is a chemical, deoxyribonucleic acid. In 1944, Oswald Avery and his colleagues proved DNA is the agent of heredity, the chemical composing genes. Nine years later, James Watson and Francis Crick showed its structure is a double helix. The chemical gene slowly emerged from obscurity.

However, Darwin raised a problem these discoveries failed to solve. Some bees, wasps, and ants survive as sterile castes in communal living systems. What could connect sterility, multiplication, and heredity? They seemed contradictory. On a simple interpretation of evolution, sterility should die out or drive the species to extinction. Darwin speculated that the evolution of sterile castes had something to do with the family. However, the biological tools to solve the problem awaited discovery until 1964.

In that year, graduate student W. D. Hamilton published two papers. In the first, he demonstrated the social effects of the unusual heredity systems of these same bees, wasps, and ants. As was well known, sexually reproducing organisms normally pass on half their genes from the mother, half from the father. Thus, offspring are 50 percent related to each parent and 50 percent related to each other. In these bees, wasps, and ants, however, sisters are 75 percent related to each other, brothers only 25 percent related. Females work, while males are drones. Their sole activity is to impregnate the reproductive queen of the next colony. Why? Due to their unusual patterns of heredity, the females pass on more copies of their genes to the next generation through their mother's reproduction than through their own. In evolutionary terms, their best option is to remain sterile, raise their sisters, and help their mother reproduce. In contrast, the males' only option is to father the next colony. Because of their genetic structure, castes of sterile female workers and inactive drones evolve easily in these species, as the theory predicts. This social structure has evolved numerous times. Hamilton's second paper applied his conclusions from these bees, ants, and wasps to the large majority of organisms that follow Mendel's rule of 50 percent relatedness.

Prior to Hamilton's publications, biologists thought of heredity in terms of survival to reproduce. The golden rule was, produce offspring! Hamilton saw another way. Help relatives! They, too, carry copies of the helper's genes. If an organism helps relatives survive to reproduce, the relatives will carry copies of its genes to succeeding generations, too. Hamilton's theory acquired two names, still in use: inclusive fitness and kin selection. Thus began the transformation in biology known as sociobiology. Sociobiology is the study of animal social behavior based on genetic relatedness. The model is simple and makes clear predictions. Closely related organisms tend to help each other, all other things being equal. The more closely related they are, the more help they tend to give.

Biologists working with natural populations immediately began to test Hamilton's model. The classic case is that of the ground squirrel. If a predator approaches a group of squirrels, one squirrel often gives a warning call. The call attracts the attention of the predator, endangering the calling animal. Before Hamilton, the sacrificial behavior of the calling squirrel appeared impossible to predict or explain. Equipped with knowledge of DNA and Hamilton's theory, biologists tested the animals for relatedness. As predicted, in groups of closely related squirrels, the warning call rang out. If a group failed to contain close relatives, silence prevailed. Biologists tested species after species where similar sacrifices appeared. In each case, relatedness provided the key to explain sacrificial social behavior. E. O. Wilson gathered the evidence from multitudes of species into one giant volume and presented it to the public in 1975.

Even the exceptions proved Hamilton correct. A few *(very few)* animals with inheritance patterns of 50 percent live in colonies with worker castes. Famous is the naked mole rat—famous, partly, for its ugliness! As the name implies, naked mole rats lack hair and, like moles, live in tunnels underground. Predictably the colonies contain only a few males or one to father the colony. Thus, the animals in the colony are unusually closely related, even by the 50 percent rule. Moreover, ecological conditions limit the available habitat. If breeding pairs abandon the colony to set up house for themselves, they cannot find a home. Therefore, they fail to reproduce, and the genes underlying their rejection of colonial living die when they do.

Another apparent exception is the Florida scrub jay. In most bird species, two parents raise their mutual offspring, building nests for them, feeding them, and protecting them. In the Florida scrub jay, extra birds assist the breeding pair. They help feed the young, protect them from snakes, and defend the breeding territory. Biologists found no convincing explanation for the helpers' behavior and could not have predicted it until they tested the birds for relatedness. As predicted by sociobiological theory, the helper helps siblings and half-siblings survive. Never do unrelated birds help.

Again predictably, ecology plays a role. As the scrub jay's name implies, the birds live in a harsh habitat. All nesting sites are occupied. Food is scarce. Predatory snakes are abundant. Young jays have no place to set up house and begin their own brood. Therefore, their best strategy for getting copies of their genes into succeeding generations is to help their close relatives. When they are older, more experienced, and previously occupied nesting sites vacated, they may begin their own families.

Of course, no insects or squirrels or birds or naked mole rats devise deliberate plans. Helpers merely pass on copies of their helping genes, and thus the next generation possesses copies of those genes and helps, also. Individuals lacking the genes fail to help, and therefore fewer of their genes survive into the next generation. Their genes diminish in number or vanish completely from the species' gene pool. Again, this is simply a case of dissimilar rates of survival and reproduction.

Sociobiology represents another growth of Darwin's theory. Its model is simple. It explained social behavior Darwin found puzzling. It made accurate predictions. It promoted research. It is taught in universities throughout the western world. It alters Darwin's phrase descent with modification of traits to read descent with modification of behavior. Kin whose social behaviors help kin tend to pass on more copies of their genes than do those who fail to help. Helping behavior among kin evolves and stabilizes, resulting in the multiplication of kin.

A major driver of sociobiological behavior is the presence of dependent offspring, especially offspring that need two parents to raise them to reproductive age. In many bird species, young birds die without the help of two parents. Several strategies evolved for the required help. Some birds appear to have evolved a communal strategy, where the community, to some extent, helps raise the young of the group. However, communal strategies are very difficult to test because they require analyzing the DNA of all the birds across succeeding seasons—no small task!

A strategy confirmed by DNA tests is for two birds to mate and remain together for a season, caring for their mutual offspring, then part. Each seeks a new mate the following season.

Another strategy proven through DNA testing is for birds to mate for life and raise their mutual young season after season. Both strategies achieve the same goal, the successful raising of young to reproductive age. However, the attachments differ. In the first, the birds appear attached to their mutual young rather than to each other. In the second, the birds appear attached to each other, to share a life-long mutual attraction. No way exists to assess their feelings (if they have feelings), but from a human perspective they appear to share a mutual love, in a birdie sort of way. Such relationships are common, occurring in about 90 percent of bird species.

Human beings produce unusually dependent offspring. Not only do our babies require complete care by the mother for several years, but mothers need the support of fathers. In our present culture, of course, single mothers may raise their children successfully. However, our ancestors spent their first millennia on the African savannas. Life there resembled a long camping trip without tents, without rain gear, without grocery stores, without medical care, and with no reprieve, while predators stalked. Single motherhood was never an option. Indeed, the isolated nuclear family perished. To survive, people lived in bands and tribes and helped one another.

However, bands are selective. Kin are the ones who bond. The sociobiology of other animals and the empirical data on contemporary bands and tribes support this contention. Logic predicts it. Two parents may be unrelated, but they are related to their children, and the children to each other. In difficult ecological conditions, whether in the Florida scrub or African savannas, families stay together for mutual aid and protection. As the children marry, they produce more individuals kin to the family. And, so, the family grows by birth into a band of some tens of related individuals, and the band into a tribe of some hundreds. All work together to support the children and each other because the families, bands, and tribes require new members as the old die and accident, disease, predation, and war threaten the young.

Abandoning the kinship group was rarely an option. To be alone was to die. To seek entrance into a band or tribe who were not one's kin courted rejection at the least, slavery in many cases, and death at the worst. That is, unless one entered as a spouse, a spouse who through mating with a tribal member would produce more kin to increase the numbers of the family, band, and tribe. Kinship promoted social behavior that, in turn, promoted the multiplication of kin and groups of kin.

Social behavior based on kinship tells a powerful story of the multiplication of kin. However, helping relationships also develop among organisms that are not kin. They practice mutual sharing. They reciprocate.

Social behavior enhances the multiplication of reciprocators

Reciprocity involves the equal sharing of goods and services. In a paper published in 1971, Robert Trivers introduced reciprocity into biology as reciprocal altruism, a term still in use. Due to the confusion among biologists and their readers about the meaning of the term altruism, however, it is best simply to call it reciprocity. Biologists, sociologists, anthropologists and mathematicians have studied reciprocity deeply. Mathematicians investigate it through game theory. The general idea in biology is to establish how reciprocity can evolve when cheating seems to offer better rewards.

Reciprocity appears throughout evolution. Mutualism, one form of reciprocity, is a major driver of evolution. It evolved among bacteria to produce nucleated cells and organelles, between fungi and algae to result in plants, and between modern plants and fungi to benefit both.

These instances of mutualism are recent discoveries. One of the first known cases of reciprocity between species is between tiny cleaner fish and their large counterparts, the fish that are cleaned. The tiny fish swim in and out of the mouths of the big fish, cleaning their teeth. What a frightful occupation! But normally the tiny fish remain unharmed. Both benefit. The tiny fish obtain dinner from the microorganisms on the big fish's teeth. The big fish gain a petite dental hygienist to keep their teeth healthy.

Initial studies of reciprocity emphasized intelligence. Clearly, reciprocity can develop most easily when the participants are able to recognize their partners, remember them, and return favors over time as well as immediately. Some very intelligent social animals appear able to do these things. Among them are our primate relatives, especially the chimpanzees. However, we are the experts. Abundant evidence of objects carried far from their places of origin tells us that bands and tribes of human beings traded with each other over extended distances long before the invention of agriculture. Contemporary tribes trade with others. Shops and customers trade, neighbors trade, nations trade. We specialize in reciprocity.

And not merely in goods and services. Reciprocity stands at the foundation of human justice systems. "Crime doesn't *pay*." "You will *pay* for your crimes." "He has *paid his debt* to society." Experiments with human subjects suggest that passion for justice lies deep within our genes. Experiments with chimpanzees and dogs indicate that these intensely social animals know when they are being treated unjustly, and they resent it. Chimpanzees and dogs that willingly perform tasks for their caretakers, but are rewarded poorly or not at all while their companion dogs and chimpanzees receive preferred treats, sulk. After repeated trials, the dogs and chimpanzees receiving the less desired rewards refuse to perform any more. They, too, seem to have a sense of justice—and injustice.

Apparently, human beings and other intelligent social animals recognize reciprocity when they experience it. And when cheated, they resent it and refuse to participate further in unequal relationships. This makes evolutionary sense. Animals that failed to recognize and expect reciprocity were cheated by their smarter and more devious counterparts. They lost resources, whereas those who cheated them gained resources. As a result, the losers survived less well, were less reproductive, and failed to pass on their genes. Meanwhile, their more intelligent and devious counterparts thrived. Those who survived best recognized reciprocity and justice and insisted on it. As experiments show, people punish wrongdoers, even at considerable cost to themselves. Moral outrage on the part of those who are treated unjustly keeps known deceivers from successfully exercising their deceptions. For this and more complex reasons, reciprocity evolves in spite of cheaters. Darwin's phrase descent with modification of traits may be enlarged again to read descent with modification of social skills, including social insight and passion. Awareness of reciprocity again led to further research, especially research on human beings. Sociobioluty is such a fertile idea, it constitutes a third synthesis.

If our social skills, including our emotions, evolved, they necessarily have a genetic base. Does this mean we are their robots, they our rulers? Are all our feelings, thoughts, and actions determined in advance by our genes? Hardly.

Determinism is a bogeyman in biology

Determinism in general is the idea that all events, including psychological events, are caused by preceding events or natural laws. Discussion goes back to the ancient Greek philosophers. Genetic determinism is a modern development of this ancient idea. The fear—it *is* a fear (count the B movies!)—is that human beings are merely cogs in a larger machine, or robots run by something-not-ourselves, programmed, like computers, by larger powers. And because the language applied to genes includes expressions like "genetic program," some biologists and philosophers see genes as the larger, external power. They think genes force us to behave (think, feel) as we do. We have no say in the matter, no choice. Others run our lives, and those others are our genes.

Often this fear is opposed by environmental determinism. According to environmental determinism, the environment determines all our actions, thoughts, and emotions. Many people find this form of determinism less frightening than genetic determinism because we have some control over our environments. In contrast, they think, we cannot alter our genes. They tend to forget that environments affect genes, so we can alter gene expression by changing our environment.

Each side tends to consider the other evil. However, a little knowledge of history shows that the evil genetic determinists (the Nazis) were at least matched in wickedness by the evil environmental determinists (the Soviet, Chinese, and Cambodian communists). Both beliefs can be dangerous. Biology substantiates neither.

Consider again the arrow leaf plant (merely a plant!). Grown in water, its leaves are narrow and flexible. A clone grown on dry land produces leaves that are broad and rigid. What happens? Environments affect genes. The watery environment turns off genes for broad, rigid leaves and turns on those for narrow, flexible ones. A dry environment does the opposite. The environment controls the genes! But not fully. Genes at the core of development always result in an arrow leaf plant. They never allow an unusual environment to produce oaks or rice or mosses from arrow leaf seeds.

Determinism in plants is hardly a philosophical issue, yet even dumb plants are neither the subjects of genetic nor environmental determinism. Genes and environments work together to produce the mature plant. This story can be repeated for billions of traits in millions of plant and animal species. Genes and environments work together to produce mature organisms that are successful in their environments. Indeed, this is how evolution works. Plants and animals respond to local environments. The successful ones survive to reproduce. The unsuccessful die. The result is descent with modification of genes and traits; descent with modification of social behaviors, capacities, and skills; and adaptation to the local environment.

Human beings have richer capacities than other animals. We possess many natural dispositions. We are disposed to like sex and sugar, pair-bonding and pornography, backrubs and bathing; to dislike insects and itches, poisons and pains, deception and defeat. The list could go on. But concerns about genetic determinism need to address the basic dispositions, those deepest in our evolutionary ancestry. For simplicity, yet comprehensiveness, four will be examined here. They are resources, reproduction, relatives, and reciprocity—easily remembered as the 4Rs.

The first R is resources. Resources are absolutely fundamental. Every living organism requires them, whether bacteria, algae, fish, fowl, or fool. Organisms need water, food, temperatures they can survive in, and (with the exception of anaerobic bacteria) oxygen. Human beings need all these things, plus safe places to allow sleep, shelter, and privacy for mating. We also require territory for our group, whether acres for hunting, pasture for herds, or arable land for agriculture. We naturally seek resources. In developed cultures, we hunt through grocery stores. We build homes with heating, air conditioning, and security devices. And, recently, we concern ourselves with air and water pollution, for we are poisoning the resources on which we depend. Seeking resources and trying to preserve them is natural and good. We need them to survive. Certainly a kind of determinism exists here. Eat or perish! provides a strong determinant.

But of what? Some people are vegans, refusing both meat and animal products. Others are vegetarians, avoiding meat, but eating eggs. Others eat meats except pork, while others exclude only beef. Some gorge on everything. Because cabbage thrives in cold climates, Russians eat cabbage. Temperate climates bring apples for the English. Warm ones grow olives for Greeks. Thus, environments rather than genes determine many food choices. But so do convictions. Vegans think it *wrong* to eat meat and animal products, even when abundantly available.

Morals and convictions inspire us. Our appetites are so underdetermined by our genes that some people go hungry and even starve to death for a cause. Status seeking, too, influences our resource choices. Everyone cannot consume caviar and champagne. Nor can every family own seven homes or mansions with empty rooms echoing. Conviction that status is good makes gluttons, while conviction that self-restraint is good forms ascetics. We need resources, but how we acquire and use them is underdetermined by our genes. We exercise a huge range of choices, from gluttony to asceticism to self-destruction.

The second R stands for reproduction. Reproduction is one of the signs of life itself. Like many organisms, we reproduce by sex. Given the amount of attention we pay to sex—moral, practical, and pleasurable—it seems to fall into the category of traits highly determined by our genes. Sex drives us. Moreover, it drives industries, from pornography to perfumes, from fiction to fashions.

And that is precisely the point. What does sex determine? When do we behave like robots? That might be answerable in some simplistic way, except that we exercise multitudes of positions for sexual intercourse. Indeed, the most noticeable attribute of our sexuality is not its robot-like character, but its creativity. We create new ways to perform it. We do it with unusual partners, sometimes with other species. We create whole industries around sex.

Of course, sexual lust drives some people's lives. No doubt about it. However, others live a celibate life—some by conviction, others because their interest in alternative activities is so great, they never quite find the time to devote to sexual pursuits or to marriage, with all its obligations.

Moreover, in most of us, sexual interest leads to love, to pair-bonding, to life-long marriages where love continues until death do us part, and beyond death for the grieving spouse.

And, of course, sex leads to reproduction, the begetting of children, grand children, and great grand children. Reproduction necessarily results in the third R, relatives.

Every reproducing organism has relatives, that is, members who share many copies of its genes. In asexual organisms and in clones, all have copies of the same genes and are 100 percent related. In the unusual sexually reproducing insects Hamilton discussed, sisters are 75 percent related and brothers 25 percent. In most other sexually reproducing organisms, siblings are 50 percent related to their parents and to each other. Genetic relationships guide social lives. Clones and asexual reproductives sacrifice themselves for their colony members, for the other members can pass along copies of the organism's genes as readily as the organism itself can, because they are all genetically alike. In select insects, sisters sacrifice their reproductive lives to raise their sisters because they are more related to each other than they would be to their own offspring. However, even the social lives of these closely related insects are somewhat more complex than these figures suggest. Some coercion occurs in the colony to keep individual females from reproducing. Some environmental pressures help colonies form and maintain themselves. Their social lives do not constitute a 100 percent genetically driven story. Rather, their extreme sociality is about 75 percent genetically driven.

In a simplistic view of sociobiology, then, our social behavior would be driven exactly 50 percent by our relationships to our nearest relatives. For example, when asked whether he would give his life to save a drowning brother, the biologist J. B. S. Haldane famously commented that he would sacrifice himself for two brothers or eight cousins. If the 50 percent rule held, we would always do so because brothers share half their genes and cousins one-eighth. But we do not. Sometimes, brothers murder one another. Sometimes they separate, never again to interact. At other times they establish businesses together. Every once in a while, twin brothers marry twin sisters. Even our closest genetic relationships result in variety, even creativity.

Yet, constants do recur. In most cases, parents love and care for their children more than they love and care for other people's children. Studies show that step children are far more likely to be abused than are biological children. Nepotism, favoritism based on kinship, is common, even though it may produce negative outcomes. Some people hire incompetent relatives or appoint them to office, much to the detriment of the institution. Nepotism has an even darker side—hatred and destruction of non-relatives to the point of genocide. No doubt special care for relatives influences our social lives. However, its influence is both complex and avoidable. Despite the ties of family, some people leave their families or tribes, never to return. They settled the New World.

Many other animals favor kin. Reciprocity, the fourth R, is the human specialty. Even so, other organisms reciprocate. As mutualism, reciprocity dates back millions of years. However, our intelligence leads us to be conscious reciprocators. We recognize individuals, remember their deeds, and return favors over time as well as instantly. We seem to have evolved a passion for fairness and justice, which are extensions of reciprocity. And sometimes our passion makes us robot-like, as when someone treated unjustly spends his or her life seeking vengeance on the perpetrator. However, most of our reciprocity is creative. We invent constitutions and institutions based on reciprocity. Reciprocity even underlies our monetary system. It is, after all, merely a sophisticated way to engage in barter, the equal exchange of goods and services.

In the case of the arrow leaf plant, genes and environments work together to produce the mature plant. Genes alone never determine all its traits. Genes and environments also work together in us, and in us, too, environments turn genes on and off. However, we are not plants. We are conscious of our surroundings, conscious of ourselves, and self-reflective. We act, and think about our actions. We think, and consider our thoughts. We reason, and examine our reasoning.

During part of the twentieth century, determinists believed brains determine thought and, if brains are computer-like, then thoughts are products of computers and therefore mere products of brain activity, lacking power to move us. Causation seemed all one-way: brains cause thoughts; thoughts do nothing. Recent work on the effects of placebos on brains has proven one-way causation wrong. Placebos are pills without active ingredients. In experiments, people are told they do have active ingredients. A sugar pill might resemble an aspirin, and the person might be told it is an aspirin and cures headaches. Swallowing the placebo often produces the expected cure.

People have long known that placebos work. This is one of the banes of the pharmaceutical industry. Often, placebos work better than expensive pills do. Recent studies using brain imaging have shown that belief in placebos changes brains. Thoughts change brains. (Faith heals!) So, do brains determine what we think? Or does what we think determine brain chemistry and structure? As with genes and environments, both are effective. This means reason and self-reflection affect brains. Reason changes us physically. It is an active cause of who we are, how we behave, and what we become.

Environments change genes. They turn genes on and off. Thoughts change brains. They activate and deactivate genes. Because we can affect our environments, we can affect our genes. Causation is multiple, rarely simple. Genetic determinism is a bogeyman. It is based on a model of gene action that is too simple to represent the world in a realistic way. Biology refutes it.

Summary

Degree of kinship affects social behavior and thus enhances the multiplication of kin. One hundred percent relatedness makes organisms genetically indistinguishable. One may as well survive as another, for each passes on copies of identical genes. Seventy-five percent relatedness usually results in the development of colonies where colony survival depends on the evolution of castes and the specialization of tasks. Fifty percent relatedness by itself determines little. Many organisms related by 50 percent never know their relatives. However, where ecological pressures demand mutual aid, close relatives help close relatives. Squirrels give warning calls. Birds bond to raise mutual offspring. So do people. Our ancestors formed large kinship groups as well, due to extreme ecological pressures. The cohesiveness of our kinship may produce negative results, leading to warfare against non kin, sometimes with genocidal results.

Reciprocity enhances the multiplication of organisms that reciprocate. Reciprocity may involve mutualism across widely separated species or trade relationships among human beings. In us, the passion for reciprocity seems to have a genetic base. We establish our economic relationships and justice systems on it.

Although genes underlie human behavior, they do not determine it. Even something as deeply ancestral as the first R, seeking resources, shows creativity in us rather than robot-like behavior. Two linked Rs also lie deep in our ancestry. They are reproduction and relatives. Sex drives our lives, but in creative ways, from varied sexual expression to exotic fashions and profitable industries. We typically prefer our relatives and form bands and tribes based on kinship. Yet, we also abandon our relatives and even murder them. Our ways of dealing with reproduction and relatives are varied and creative. The same is true of the fourth R, reciprocity. In reciprocating, we are perhaps at our most creative. Genetic determinism is a bogeyman, not a fact.

Because almost everything in biology has exceptions, nearly everything in this chapter has exceptions.

5: SYNTHESIZING BIOLOGY

Thus far, evolutionary biology has undertaken three syntheses: Darwin's, the New Synthesis, and sociobiology. Historians have researched Darwin's work so thoroughly they call their labors the Darwin Industry. Knowledge of the history of the two other syntheses is less developed, less familiar, and lacks perspective. They occurred too recently. Nonetheless, this seems a good place to pause to examine the three syntheses as science. Do they fit the criteria of creating simplified models of the world, of forming new syntheses, of making accurate predictions, of provoking new questions, and of establishing new research traditions that result in fresh publications and institutions? For Darwin, these questions have thorough answers. However, only a brief sketch is possible here.

Darwin creates the primary synthesis

It would be easy to call the three syntheses, synthesis one, two, and three. However, Darwin's synthesis is foundational. He began modern biology. He proposed the crucial ideas. Therefore, it is perhaps best to distinguish Darwin's synthesis from the others by calling it the primary synthesis—primary because it was first, but also because it is fundamental, central, and indispensable. It is the basic model that the other syntheses not only copy, but also use.

The word biology was coined in 1800 in German and came into English in 1819. There were no professional biologists until after the publication of *On the Origin of Species*. Darwin and his predecessors were naturalists—they investigated the natural world. Most were amateurs and many were clergy. They were collectors and explorers. They amassed data. They gave the data some coherence. Carolus Linnaeus organized the classification of organisms between 1735 and 1758 and proposed the two-name system we still use for biological classification. We are *Homo sapiens.* The common dog is *Canis familiaris.* The domesticated cat is *Felis catus.* Its fierce African cousin of the roaring mane is *Panthera leo.* Before Darwin's time, naturalists realized living forms fell into groups within groups. However, they did not know why organisms are arranged this way.

Geologists, too, were collectors and explorers. By Darwin's day, they realized geological strata are orderly. The younger lie over the older. Moreover, all but the oldest strata contain fossils, and the fossils distinguish one stratum from another. But why this is so constituted a mystery.

Fossil hunters explored the globe. They knew that fossils, in general, went from very simple in older rocks to more complex in younger ones. They knew that the most recent fossils tended to resemble animals living in the vicinity. That is, the fossils in Africa resembled African animals. In Australia, the fossils resembled Australian animals. And so on. As they explored deeper strata, the fossils resembled living animals less and less. Why these things are so, no one knew.

Exploration of the distribution of living organisms and fossils across the globe constitutes biogeography. Biogeography raised many questions. An early question was how animals in the Americas and Australia traveled to and from the ark, which the Bible said Noah constructed in the Middle East. By the middle of the eighteenth century, Linnaeus and others who monitored the discovery of new species realized that no boat could contain all the known species. However, that only raised new questions. Were the animals of the African desert distinctly African, or were they typical desert animals whose species might also exist in unknown deserts? The same species of plant lived in the arctic and on the mountain tops of Europe, but none lived between the two frigid areas. How could the same immobile species arrive in two geographically separated places? Were fossil creatures really extinct, or had their living counterparts simply migrated to unexplored regions of the globe? No one had answers.

Anatomists, too, discovered patterns. The limbs of four-legged animals were similar. They had one bone near the body, two farther out, followed by a series of smaller bones, followed by digits (toes, fingers). Bat wings had the same structure. So did bird wings. So did human arms and legs. There were other similarities, too. No one knew why.

Some organs in animals seemed undeveloped. Others, like the human appendix, seemed useless. Others appeared less well-designed than an engineer might make, like the human back, so prone to injury. Why, people wondered, were organisms less perfect than they might be?

Tiny embryos in distinct species appeared remarkably similar. However, as they grew larger, they began to resemble each other less and less. In species so unlike, why would embryos resemble each other? No one knew.

Not only did Darwin's theory of evolution—descent with modification—explain all these cases, it tied them together. For the first time in history, they made sense both in themselves and in relationship to each other. One explanation, one theory, one simple model of evolution, created a huge synthesis of the known facts of biology and geology.

Darwin's simple model showed that organisms change. It predicted their evolution. It showed that one species is ancestral to another. Moreover species go extinct. The theory of evolution explained why fossils in similar strata are similar: they lived at the same time. Extinctions explained gaps between them, as did missing strata in the geological record. Living organisms descended from ancestors that had become fossils. Recent fossils resembled living forms while earlier ones differed from them because change occurred through time: the longer the time that passed, the more distant the ancestor, and the more change occurred. Evolution rested on these discoveries, but it also predicted them. The predictions proved accurate. Everywhere the fossils changed over time, becoming more like living organisms as they rose higher in geological strata.

Mammals resembled each other in the anatomy of their limbs, which was predictable from evolution because they inherited the patterns from common ancestors. The pattern repeated in the wings of bats because they, too, were mammals, sharing common ancestors. The wings on birds displayed a similar pattern. Yet, because they possessed fewer common traits with mammals than mammals did with each other, their common ancestor must have been more distant than that of the mammals.

Biogeography was clarified. Organisms were African, Asian, European, Australian, or North or South American because impassible or nearly impassible barriers isolated continents from one another. Organisms originating in Africa stayed in Africa and evolved there, changing over time, just as the African fossils showed. The theory was predictive. The same proved true for every continent. Moreover, islands carried species that lived nowhere else because their ancestors arrived there from the mainland, became isolated, and then evolved into new species.

Evolution—descent with modification—explained why living organisms fell into groups within groups. It predicted the pattern. The groups that shared many traits had a near common ancestral population. Those that shared fewer traits had a more distant one. Moreover, common ancestry explained many difficulties in classification. One of the most frustrating was distinguishing genuine species from varieties within species. Evolution said varieties might constitute emerging species! Over time, they might change enough from the parent species to become genuine species themselves. How? By natural selection. If offspring varied (they did), if they inherited some variations (they did), if more offspring multiplied than survived to reproduce (they did), then populations would change with each generation! Over time—over very long periods of time—they might change just as the fossil record indicated. Geological estimates based on the slow deposits of sedimentary rocks and the slow erosion of volcanoes provided the time. Darwin, whose first discoveries were in geology, predicted Earth to be more than a billion years old. (Decades later, scientists confirmed his prediction of an old Earth with the astounding discovery that Earth is more than four billion years old. The ability to date rocks gave evolution a time line and made fossil discoveries predictable.)

Darwin's theory of descent with modification was remarkable. It answered dozens of questions. It made accurate predictions. It synthesized separate topics. The first edition of the *Origin* sold out the day it was printed. Everyone read it. During Darwin's lifetime, it went through five more editions. Scientists and non-scientists alike accepted evolution. Darwin provided them with a new world view that made sense of the evidence. Like Newton, he offered a natural explanation, one that referred to causes that could be tested by observation and experiment, then confirmed or refuted. Indeed, Darwin's theory had already been tested. It was a synthesis of centuries of observations, of ancient and current domestic breeding programs, of carefully designed experiments. Accepting it required only simple logic and releasing former ideas about how organisms come to be. Some people balked. But many found Darwin's theory liberating. Natural causes could explain natural events, after all! Darwin's work constituted an extension of the European Enlightenment. And the European Enlightenment owed much of its light to Newton.

Darwin's mechanism of natural selection was not accepted so easily. It seemed too chaotic, too dependent on chance. The British scientist, William Hershel, called it the "law of higgledy-piggeldy." As a result, it motivated little research until the twentieth century.

In contrast, Darwin revolutionized fossil studies. No longer were fossils merely successions of bones in rocks. They represented the history of life on Earth. Almost immediately, scientists began to recreate that history. In the process, they accumulated additional evidence for evolution. They established almost the entire history of the horse. The series of fossils began with the little, four-toed eohippus, branched remarkably (more bush than tree), and resulted in the large, hoofed animals known to all. Cartoons appeared of tiny eohomo astride tiny eohippus.

Other gaps filled in. People discovered predictable transitions. The transitional reptile-to-bird fossil, *Archaeopteryx,* came to light immediately, in 1859. Lungfish had been known since the 1830s. Now biologists realized they might be transition species: fish, yes, but with lungs like land animals. Much research ensued, with many fish and amphibians surveyed and investigated as transition animals. Learned papers and books appeared. Darwin's ideas on evolution inspired an enormous amount of research on this transition.

And on other transitions. Biologists sought insect origins, the lineage of the crustaceans, the ancestors of worms, fish, amphibians, marsupials, and mammals. Darwin inspired all these researches.

Evolution revolutionized classification. Now it expressed biological relationships of ancestry and descent. This helped clarify a persistent difficulty. Some animals resembled each other in shape. Others were alike in traits that were less obvious but known to be more important for classification. Now biologists understood that evolutionary adaptations to similar environments could result in similar traits without indicating ancestry. Thus, whales resembled fish in shape because they both moved through water, and efficiency evolved a bullet-like shape. Evolutionary adaptations would produce similar shapes without inheritance from a common ancestral population. Using such traits for classification was an error—an error because Darwin showed that the long-sought natural classification was classification of ancestry rather than of similarity.

The synthesis Darwin provided also resulted in interdisciplinary programs in evolutionary biology. Biologists saw that fossil studies, biogeography (which included fossils as well as living organisms), and anatomy were related. The fields exchanged information, enlightening one another. Thus, Darwin's synthesis led to a synthesis of disciplines.

The profession of biologist attracted recognition and funds. Museums had housed professionals dealing with classification and anatomy since the eighteenth century. However, most of the work in biology was done by amateurs until the 1850s. Educational systems enlarged during the late nineteenth century. As they did, professorships expanded everywhere, and biology appeared among the expanding fields. The first marine biological stations were established beginning in 1872. Governments, universities, and museums began funding expeditions. The voyage of the H. M. S. *Challenger* from 1872-1876 was the first of many. American universities, in particular, began funding biological laboratories which were to play such a prominent role in twentieth century science.

Established journals like the *Philosophical Transactions of the Royal Society, Proceedings of the Zoological Society,* and the *Quarterly Journal of the Geological Society of London* accepted papers related to evolution. New journals sprang into being. Britain saw the *Quarterly Journal of Microscopical Science.* In America, the *Journal of Morphology* appeared. Semipopular publications like *Nature,* the journal, began publication and were widely read. Rarely has one simple theory proved so productive.

Professionalization and professional publications devoted to biology in general and evolution in particular increased in the twentieth century. Many of these anticipated the second synthesis.

It was clear to everyone including Darwin that the central problem of his theory was heredity. Early and continuing efforts focused on the physical carriers we now call genes. Scientists proposed a dozen ideas before 1900, but each was shown to be flawed. The difficulty with discovering the physical carriers was technical as well as conceptual. Technically, the heredity material was microscopic, and means of viewing it proved inadequate until the 1950s. Conceptually, science had to refute the ideas of blending inheritance and the inheritance of acquired characteristics before the material could be understood well enough to be recognized if found.

Concentration on the problem of the heredity material only increased with the discovery of Mendelian genetics in 1900. In hindsight, however, the discoveries and speculations between then and mid-century proved either insignificant or irrelevant or both. Only in 1944 did biologists discover chemical genes as DNA and in 1953 the structure of DNA as a double helix. Thus, modern knowledge of chemical genes occurred only after the completion of the second synthesis.

The breakthrough Mendelian genetics provided helped change the focus of research on heredity. Mendelian genetics allowed biologists to ignore the unknown material carrier. Darwin's theory of evolution, after all, is a theory about changes in populations. It is a statistical theory. Therefore, Mendel seemed to whisper, do the statistics. Do the statistics for the transmission of genes in static populations (the Hardy-Weinberg law for stable populations). Do the statistics for the transmission of genes in changing populations (population genetics). Do the statistics for the transmission of genes down the generations (transmission genetics).

So mathematicians did the statistics. They were joined by experimentalists who devised creative techniques to answer statistical questions and test statistical predictions. Meanwhile, biologists in the older tradition of Darwin continued their observational field-work. By the 1930s, the two research traditions separated. They spoke different languages, asked different questions, and explored different conceptions of biology. Each had much to offer the other. However, each also had to discard misconceptions that arose from narrow specialization. At the annual meeting of the Geological Society of America in 1941, Walter Herman Bucher proposed a synthesis. Biologists responded. Concerned geneticists, fossil specialists, and classification professionals met in 1942 at Columbia University and agreed to become more familiar with each other's techniques and discoveries. The Second World War interrupted their meetings, but not their efforts. In 1947, an international symposium met at Princeton, New Jersey, with representatives from almost all fields of biology. They agreed on significant matters: evolution is gradual, natural selection is central, diversity occurs through the evolution of populations. All also agreed that the synthesis began with the publication in 1937 of *Genetics and the Origin of Species* by Theodosius Dobzhansky who first demonstrated the fit between Mendelian genetics and evolution.

This was not a revolution. It was a synthesis. It rid biology of accumulated errors: the inheritance of acquired characteristics, abrupt evolutionary transformations, and all types of vitalism that claimed evolution is driven by mysterious internal forces in the organism rather than by the external action of natural selection. It established natural selection as the main driver of evolution, vindicating Darwin's unique contribution to the theory of evolution. It rightly gave mathematics an important role, for evolution occurs in populations and therefore requires statistical analysis. It furnished the standard approach to evolutionary biology for more than half a century.

It inspired the founding of new journals. Especially notable was the appearance of *Evolution,* the journal founded by Ernst Mayr. This journal continues to be a major avenue for the publication of professional papers in evolutionary biology.

The second synthesis also opened up new vistas for research. Some were conceptual. A new discipline arose: philosophy of biology, usually housed in philosophy departments. The discipline asked such questions as, What is the structure of the theory of evolution? What is a law in biology? What is a species? What is a gene in Mendelian genetics? What is the relationship between the mathematical gene and the chemical gene? What is the connection between genotype and phenotype? Early works appeared by the philosopher David Hull *(Philosophy of Biological Science,* 1974), and a philosopher who later also became a historian of biology, Michael Ruse *(The Philosophy of Biology,* 1973). Ernst Mayr (of course!) contributed a mammoth 564-page study *(Toward a New Philosophy of Biology,* 1988). Philosophical books continue to appear. Universities offer courses.

Other vistas were experimental and observational: to describe in detail the workings of selection, the role of isolation in speciation, the production of variation, and the role of randomness in evolution. Especially notable was Hampton Carson's work on speciation in the fruit fly on the Hawaiian Islands, with their numerous isolating elements including oceans, streams, volcanoes, and lava flows. His work confirmed Darwin's simple model, expanded and publicized by Mayr, of isolation as a major cause of speciation. It also addressed issues of selection and randomness and inspired further work on speciation in Hawaii and elsewhere.

This research and much more fell fully within the second synthesis. Exploration of the social behavior of animals did not. Rather, it returned to Darwin's insights, expanded them, and filled in details through the development of theories, observations, and experiments.

The third synthesis fulfils Darwinian initiatives

The second synthesis treated genes as if they were individual atoms, unconnected to each other. The third synthesis took the opposite approach. It studied genetic effects collectively. It began with one of Darwin's basic insights: evolution is about survival to reproduce. It then asked why some social animals seemed to behave in a self-sacrificial manner. Some animals risked their lives to warn others in their group of danger. Others sacrificed their reproductive opportunities as well as their lives to feed and protect members of their group. Darwin predicted that the answer lay with relationships within the family. However, he lacked the information to show why or to provide details.

A century later, biologists knew the requisite details of heredity. Usually, parents share 50 percent of their genes with their mutual offspring, and the offspring share 50 percent of their genes with each other. Some ants, bees, and wasps constitute a special case in which sisters share 75 percent of their genes, while brothers share only 25 percent. Sociobiology made Darwin's answer quantitative and provided precise predictions: the more closely related animals are to each other, the more sacrifices they will make for each other. The answer Haldane gave to the question of whether he would give his life for a drowning brother proved spot on! He would save two brothers or eight cousins. Why? He is half related to his brothers and one-eighth to his cousins.

The quantitative answer inspired decades of field research that biologists are still pursuing. In species after species, whether mammal, bird, or insect, the quantitative predictions proved correct. Of course, there were qualifiers for animals less than 100 percent related. Environment influences helping behavior, too. A difficult environment forces animals that might not otherwise cooperate to cooperate or die. But the environmental influence merely enriches and strengthens sociobiology. It does not refute its basic theory that the more closely related animals are, the more they help each other.

Sociobiologists founded new journals like *Ethology and Sociobiology*. As some sociobiologists increasingly applied their insights to human beings, they founded journals related to biology and human social behavior. An important journal is *Evolution and Human Behavior*. Books appeared in both fields: *The Triumph of Sociobiology* by John Alcock proved readable for non biologists. David M. Buss produced the standard textbook in evolutionary psychology, *Evolutionary Psychology: The New Science of the Mind*. University courses multiplied.

The other Darwinian initiative that the third synthesis fulfilled was sexual selection. Darwin pursued the subject in a book-length work. He laid the groundwork. For a century, however, his insights lay unheeded. Because sociobiology focused on social relationships, and among non-social animals social relationships center on mating, sociobiology stimulated research on sexual selection. The research showed Darwin's predictions correct: females do select mates carefully, while males show far less caution in mating. However, Darwin with his limited knowledge could not offer sound reasons, despite his meticulous and extensive observations. Discoveries in biology in the twentieth century enabled biologists to provide reasons. The reasons, again, are more or less quantifiable and make simple predictions. In a word, females expend considerable energy bearing each offspring and are limited in the number they can bear, whereas males expend little energy fathering each offspring and can produce unlimited numbers. Because she is limited, the female must make careful choices about a mate if her offspring are to multiply down the generations. The male is far less limited, so he can afford to mate with every female he finds (if she will have him!), whether she is a sturdy reproducer or not.

Darwin initiated the idea and offered a multitude of observations to support it. Sociobiology supplied the theory. Modern research substantiated both. Recent books include James L. Gould, *Sexual Selection: Mate Choice and Courtship in Nature* and Peter M. Kappeler, *Sexual Selection in Primates: New and Comparative Studies.*

Reciprocity is another area the third synthesis pursued. So much research has occurred outside biology proper, both before and after the initial paper by Robert Trivers, that little can be said here. Perhaps it is sufficient to note that the mathematics of game theory and the insights of biology, when synthesized into a single model, have produced remarkable results. An insightful and prolific author is John Maynard Smith.

Summary

It has been said that all of western philosophy is but footnotes to Plato. Likewise, it might be said that all of modern biology is but footnotes to Darwin. In neither case, of course, is this quite fair. However, Darwin most certainly developed the primary theory behind all modern biology: evolution through natural selection— and sexual selection. He produced a remarkable synthesis of biological knowledge available during the last half of the nineteenth century, and his work has inspired research in biology ever since.

Seen from a theoretical perspective, subsequent syntheses merely resynthesized Darwin whenever new discoveries in biology threatened to tear apart his original synthesis. Genetics became separated from observational field work. The second synthesis rejoined them while also affirming Darwin's original insights. Modern biology quantified Darwin's verbal theory: mathematicians applied Mendelian genetics to explain heredity through gene flow within, across, and down populations. Sociobiology employed mathematics to family relationships, to sexual selection, and to reciprocal relationships. The new discoveries and quantifications supported Darwin's original model, refined its predictions, expanded its range, and enhanced its details.

Nonetheless, modern biology has failed to resynthesize Darwin's original synthesis completely. Darwin's synthesis included embryology as one of its essential proofs of evolution. The second synthesis excluded embryology. The third synthesis ignored it: its focus was elsewhere. Moreover, discoveries in molecular biology—insights about the chemical gene—remain outside all the syntheses. Darwin knew nothing about genes. The second synthesis concentrated on the mathematics of gene flow and ignored the chemistry. Moreover, biologists completed the second synthesis before they understood the structure of DNA. For the most part, the third synthesis focused on social relationships, mentioning chemical genes only when they were immediately relevant. When it included genes, they were often of the one-gene, one-trait, "genes for" variety. Current models of development focus on chemical genes. They include embryology. Biologists are making astounding discoveries.

Thousands of studies are available that substantiate and fill in the sketch presented here.

6: APPRECIATING DEVELOPMENT

Beginning in the 1980s, the chemical gene came into its own in developmental biology. Soon, biologists applied the accumulating knowledge to evolution. They established a new discipline, evolutionary developmental biology, dubbed evo-devo. The founding text appeared in 1996 and ran 520 pages. It is *The Shape of Life: Genes, Development, and the Evolution of Animal Form* by Rudolf A. Raff. Discoveries and considerations of them flow in steadily. Two clear and helpful newer books are *Endless Forms Most Beautiful: The New Science of Evo-devo and the Making of the Animal Kingdom* by Sean B. Carroll and *The Plausibility of Life: Resolving Darwin's Dilemma* by Marc W. Kirschner and John C. Gerhart. The breakthroughs of evo-devo are astounding. Most remarkable of all is the discovery that the same core genes and processes are at work throughout the organic world.

Core processes allow variation

The second synthesis of the 1930s and 1940s rested on the idea of descent with modification of gene frequencies. The mathematics assumed one gene, one trait, although everyone knew this was a simplification. The mathematical treatment of genes implied that complexity multiplied as the number of genes increased, and variation depended on the origin of new genes. The best biologists thought eyes had evolved anew some 40 to 60 times as novel genes for capturing light appeared in a variety of species.

Recent discoveries have disproved these assumptions. The disproof required detailed knowledge of embryology, now called developmental biology. Developmental biology was slow to mature. There were reasons. Development depends on genes. But genes are difficult to investigate. They are both complex and microscopic. Even after biologists understood their chemistry and unraveled the structure of DNA, observing their role in development proved difficult because biologists are unable to use the usual technique of chopping organisms into pieces and studying the pieces. They must observe the role of genes in development in whole, living organisms. Until the 1970s, techniques for doing so were few. When techniques improved, discoveries flooded in. Now biologists are trying to synthesize developmental biology with evolution. The enterprise is evolutionary developmental biology, evo-devo.

Only as evo-devo advanced did it prove the assumptions of mathematical genetics false. The decisive event was the discovery of the *Hox* genes in the 1980s through intricate laboratory work by biologists world-wide. The *Hox* genes are a group of genes found in all bilaterally symmetrical animals. And almost all animals are bilaterally symmetrical, that is, the left and right sides form mirror images of one another. We have right and left arms, right and left hands—legs, feet, eyes, etc. The *Hox* genes regulate this bodily symmetry and how it makes different appendages different. They appear in the same order on the chromosomes of each species, laid out from head to tail. Moreover, they turn on in the same order in each species, also from head to tail. The only reasonable explanation in evolution for the various animal species having the same genes (and, moreover, in the same order) is that the genes passed down from a common ancestral population. This common ancestral population must have evolved a long time back. Biologists estimate that its appearance predates the Cambrian Period of 500 million years ago, when most animal forms appear for the first time in the fossil record.

The *Hox* genes are regulatory genes that stand at the very head of cascades of genetic activity. They are master genes. Although they are the same in all animals, they produce different animal forms because switches turn them on and off in different locations, in different times, and in different contexts, depending on the animal.

Soon biologists discovered other master genes. *Eyeless* (later *Pax-6)* controls the development of eyes. It was first called *Eyeless* because without it, no eyes develop, despite the presence of other eye-forming genes. (By convention, the names of genes are in italics, whereas the names of their proteins are not. Thus we have the *Pax-6* gene and its Pax-6 protein.) *Distal-less* (later *Dll)* controls the development of appendages (those protrusions that grow away—distal—from the body). Without it, no appendages develop, despite the presence of other appendage-forming genes. *Tinman* (later *Nk2),* named after the tin character who lacks a heart in *The Wizard of Oz,* similarly controls the development of hearts. All these genes are ancient. All control core processes.

Thus, by the beginning of the twenty-first century, biologists began to understand evolution as a conservative process as well as a creative one. It conserves core genes and core developmental processes. Moreover, there are no "genes for" worms or "genes for" cats or "genes for" human beings. These models were too simple. They confused genotype with phenotype. Biologists now know that the same genes appear in all organisms. However, they combine differently. Consider by analogy a deck of cards. There are only 52 cards, but we play thousands of different games with them. The different games merely require different rules, not different cards. In organisms, genes are like cards. They remain largely the same throughout evolution. Different rules for combining them create the differences among individuals, species, and even larger biological groups, like phyla or even kingdoms.

Like many analogies, this one overlooks details. Complex organisms do display more genes than simple ones do, and some of their genes are unique. Bacteria and fungi lack *Hox* genes, *Pax-6, Dll,* and *Nk2.* These genes appear in their present form only in animals. Yet, animals and bacteria share many genes. Some knowledge of the story of the evolution of core genes and processes as biologists now understand it may be helpful.

Bacteria constitute the simplest organisms. They evolved first. They are prokaryotes, single cells without a nucleus. They manipulate DNA and its simpler cousin RNA, code for the 20 amino acids that construct the proteins in all organisms, form membranes that divide inner from outer, synthesize sugars, and multiply. Because all organisms are composed of cells, and all cells do these things, genes for these core processes appear in all organisms, including human beings.

Eukaryotes evolved next. They are single cells that contain a nucleus, an internal structure, and organelles. Additional genes that underlie the formation these things along with the processes that maintain them exist in eukaryotes. Because all multicellular organisms are composed of eukaryotic cells, all multicellular organisms possess these genes. In addition, they have genes to control signaling among cells and to produce the binding material that enables cells to stick together.

All these genes exist in complex organisms. Animals have additional genes. Consider the first bilaterally symmetrical animals.

Biologists dub them the urbilaterians, meaning the first or original or primitive bilateral animals. Because the urbilaterians were composed of eukaryotic cells, were multicellular, and were bilateral, they possessed the genes in eukaryotic cells, those for multicellularity, and those for bilateral symmetry, the *Hox* genes, which turned on from head to tail. In addition, they had genes for animal features: *Pax-6* for eyes, *Nk2* for hearts, and *Dll* for appendages. They possessed a number of different cell types. They had muscle cells, nerve cells, photoreceptive cells, and digestive cells, among others. All these cell types, their genes, and their core processes date to more than 500 million years ago.

Almost three billion years passed from the origin of life until these core genes and processes were in place. After that, evolution got easier. It had two jobs: maintain the core genes and processes, then add, duplicate, and specialize. Eyes and hearts could become more complex. Segments could be multiplied and specialized. Appendages could become legs, antenna, wings, claws, hooves, and hands. Francois Jacob wrote more accurately than he knew when in 1977 he referred to evolution as a tinkerer rather than an engineer. Evolution does not create new creatures out of nothing. It creates them from used parts.

Having the core processes in place allows variation. The core processes provide for stability and viability. Organisms whose core processes fail, die. Those whose core processes thrive, survive to reproduce. Thus, natural selection conserves organisms with viable core processes. Such organisms can change through changes in regulation, yet retain viability. Natural selection can diversify their anatomy, evolving wings from legs. It can alter their physiology, creating new tissues, organs, and cell types. Natural selection can improve their vision based on the simple light receptors provided by *Pax-6* and evolve true hearts from primitive *Nk2*-derived pumps. It can create a multitude of appendages with many functions. The main source for these creative changes is duplication.

Duplication supplies variation

The evolution of development has a theme: duplicate, then specialize. Many parts of organisms may be duplicated, from genes and chromosomes to cells, organs, segments, and appendages. The duplicated part is often co-opted, that is, changed from one function to another. Insects duplicated a bodily segment then, on the new segment, they co-opted the old large wings for propelling flight into blunt wings for stabilizing flight. Evolution can proceed easily in this way because the original part continues to perform its old function, while the duplicated part specializes to perform a new one.

The original ancestors, the urbilaterians, had only one *Hox* cluster. Early in evolution, the *Hox* cluster duplicated, and animals continue to exist today with only two *Hox* clusters. By the evolution of the early vertebrates, the *Hox* clusters duplicated again. As different as today's vertebrates appear, all amphibians, reptiles, birds, and mammals have four *Hox* clusters. The difference in their use lies initially in their location in the embryo.

Arthropods include insects like bees and flies, and crustaceans like lobsters and crayfish. Their evolution is fairly well understood because fruit flies are a favorite laboratory subject and uncontroversial. Arthropods evolved by duplicating their segments along with the segments' appendages. Evolution continued as the appendages specialized through co-option for a multitude of functions, from feeding to locomotion, from respiration to defense.

In contrast, eye evolution is still misunderstood, partly because it is controversial. The question dates to before Darwin. Back then, the eye was the favorite example of an organ that could only be explained by divine creation. People argued that it must come into existence as a whole. What good, after all, is part of an eye? And what could form it whole but a divine, intelligent designer? Darwin's intimate familiarity with animals like flatworms with their simple eyespots led him to suggest that any light-capturing organ is of more use than none at all and, once formed, it might increase in usefulness as it evolved greater complexity. Conversant with evolution, the second synthesis thought eyes evolved anew some 40 to 60 times because there seem to be 40 to 60 kinds of eyes across the animal kingdom.

Increased knowledge of the chemistry of genes led biologists to realize that genes discovered in a variety of organisms and named differently were in reality the same gene. *Eyeless* in insects is the same gene known in mice as *Small eye* and in human beings as *Aniridia*. When biologists replaced

Eyeless from an insect with *Small eye* from a mouse, insect eyes developed in the altered insect, not mouse eyes. This experiment proved beyond a doubt that *Eyeless* and *Small eye* really are the same gene. The same gene forms different types of eyes in different contexts. Biologists then eliminated the confusion of names for the same gene by giving it a common name. It became *Pax-6* in all species. They discovered that this gene produces eyes even if it is relocated to limbs. It is a master gene. As such, it must date back to the urbilaterians. The urbilaterians' primitive eyes needed only two cells: one to detect light and the other a pigment to control the light's angle. Similar simple eyes exist today in various larvae.

With this knowledge in hand, biologists realized for the first time how easily so many different kinds of eyes could evolve. The evolution of eyes followed the basic theme of the evolution of development: duplicate, then specialize. Keep the core processes stemming from *Pax-6,* then duplicate the cell types and reorganize them to create eyes specialized across the animal spectrum, depending on the environment. Evolution forms the diversity of eyes from old parts. It is a tinkerer, reorganizing ancient parts to make new eyes, rather than an engineer inventing each kind of eye fresh from nothing.

The theme of duplicate, then specialize, repeats itself in the evolution of limbs from the fins of fish. Not any type of fin can transform into a limb. Most cannot, for they are fan-shaped, and it is difficult to turn a wide fan into a focused limb. However, the ancestors of lungfish that existed some 365 million years ago had centralized bones in their fins. The difference between their swimming fins and later walking limbs is small. In the fish, genes turn on only along the back of the fin. In limbs, genes turn on across the front of the limb also. This is a limb without feet. However, a simple *Hox* switch in a new place creates the bones of ankle/wrist and toes/fingers with the repeated use of the chemical alanine. The more alanine, the more digits. Indeed the ancestors of modern animals had eight digits on their front feet and seven on their rear feet. However, there were only five types of digit, and evolution eventually reduced the amount of alanine and therefore the number of digits. Today, five is the standard number, although some are fused or reduced to produce hooves.

Evolution continued. Vertebrates that walked on land found a whole new niche to invade—the air. The trick was to turn front limbs into wings. Evolution performed the trick three separate times in three different ways. Birds have arm wings; pterosaurs, finger wings; and bats, hand wings. In birds, the feathers of the wing grow all along the forearm. The hand remains short. Pterosaurs use the arm, but extend the forth finger of the hand to support a membranous wing. Bats employ both arm and hand, but the membrane of their wing attaches to fingers two through five, which are quite long, then stretches all the way to the heel of the hind limb.

In contrast to walking and flying creatures that transformed existing appendages, snakes discarded theirs. *Hox* genes, both front and back, ceased to be expressed or were expressed only as tiny buds. Instead, the snake duplicated its spine and ribs, then specialized them for locomotion.

Animals are modular. They have core functions that cannot be disrupted without causing death. But everything downstream from the core can be duplicated, then specialized for new functions. Downstream changes leave most of an animal unchanged, although some rearrangement may be required. However, the genes for core processes remain the same. The changes may add or delete genes, but usually they duplicate items, then specialize the duplicates. Such changes may result in new species, even higher categories, without adding important new genes. We now understand how duplication supplies the variation for natural selection to work on. Natural selection uses duplicated parts, leaving the old parts to perform their necessary functions while specializing the new ones for new functions. Changes in embryo development underlie this process.

The embryo provides the key

The embryo begins as one cell, created from the union of sperm and egg. This single cell contains a copy of the entire DNA for the organism. It is unique because it mixes half its DNA from its mother with half from its father rather than receiving its DNA from a single source. Nothing quite like it has existed before. The DNA in this one cell is its genotype. In contrast, the phenotype is the whole organism, with all its traits. Because the phenotype is mostly protein and water, and DNA is neither, the phenotype cannot be the unmediated product of DNA, which is an inactive chemical. The genotype is not a blueprint, containing all the information for building the phenotype. The phenotype contains much information lacking in the genotype.

The developing embryo uses the same theme that the evolution of development employs: duplicate, then specialize. Almost as soon as the original single cell forms, it begins dividing. Its new cells duplicate repeatedly, with each cell continuing to contain a copy of the entire DNA for the organism. The initial cells (stem cells) are able to become any type of cell. Their DNA trembles with potential, just waiting to be used. Before long, the chemical context around cells starts them on the road to specialization, providing variation. The chemicals in the surrounding matrix turn some genes on in these cells, some off. And some of the genes that are turned off remain off permanently. If the cells have become bone cells, they can no longer be reconfigured to become muscle cells. If they have become skin cells, they can no longer become lung cells or any other kind of cell. They have specialized, and they cannot turn back. They inherit certain information by epigenesis, by initial influences on the genes rather from the genes themselves. The new information is not in the DNA. All changes in the cells, their movement from one place to another, their specialization, and their adhesion to one another, occurs in context. Development is guided locally. There is neither command center nor blueprint.

In its initial phases, the context of the embryo is the same for all bilateral animals. Every embryo develops a simple geography: head-tail, bottom-top, left-right. The cells in each of these areas begin to specialize. Cells in the head region begin to specialize as head-parts. In the tail, they become tail-parts, etc. Cells that secrete chemicals make the specialized divisions. The cells within particular divisions respond to the chemical gradients around them. The chemical context turns certain genes on and off, so the cells become head or tail, top or bottom, left or right. This simple geographical specialization occurs in all bilateral animal embryos. It is the central core of the core processes. It is controlled by *Hox* genes. If it goes wrong, the embryo dies as a natural abortion.

When the embryo successfully completes these simple geographical divisions, it makes more. Soon, it has created specialized compartments. In the head area, compartments appear where specialized brain parts develop: the hind brain, mid brain, and forebrain. Other compartments divide areas for the spine and internal organs. Appendages begin as compartments within compartments that sprout three-dimensional buds. These buds also develop geographically, from the body outward. As they extend outward, they, too, specialize. Thus, the upper arm bone develops first, then the lower arm bones, then the wrist, then the digits. Here *Distal-less (Dll)* is busy. It sends cascades of signals down the arm as it develops. In snakes, *Dll* lies silent, or hiccups briefly and dies.

Not only is development guided locally rather than by central command, but some of it depends on explorer cells with the potential of stem cells. They randomly explore an area until they receive signals or arrive among chemical conditions that activate their appropriate genes. These cells include those for the smaller nerves, muscles, and blood vessels, and the immune system. Lack of oxygen, for example, turns on genes in cells that then specialize to make new blood vessels that will supply the area with oxygen. Cells whose exploratory behavior fails to place them an appropriate environment die. In this manner, Darwinian processes of dissimilar rates of survival and reproduction occur within the organism itself and guide its development.

Explorer cells facilitate evolution. They also answer questions about irreducible complexity. When appendage bones change, as in fin to limb, or mature differently, as in horses and people, no genetic change is required to develop the smaller nerves, muscles, and blood vessels in the places necessary for the smooth functioning of the organism. Bone development is fairly well guided by cascades of Dll proteins induced by *Dll* genes responding to the context of appendage development. However, the nerve, muscle, and blood vessel cells explore until they arrive in the appropriate context for development, then multiply. If they fail to settle in the appropriate context, they die. The contexts differ as the bones differ. So explorer cells develop according to the requirements of the organism without changes in genes and without an overall plan—or even a local one.

The response of cells to context is familiar to biologists and can be seen by anyone through a simple experiment. Place stem cells (cells with potential to become any specialized cell) on different gels, some soft, some intermediate, some hard. Stem cells develop into nerve cells on the soft gels, into muscle cells on the intermediate gels, and into bone cells on the hard gels. The experiment shows beyond doubt that development really is local and contextual, without an overall guide.

Sensitivity to context requires that changes of context result in changes of form. If the location of *Hox* genes changes in the embryo, the result is an alteration of animal form. Change of animal form often results in speciation. The creation of new genes, then, is unnecessary for the evolution of new species. Knowing now that animals share many genes across species (and organisms across kingdoms), this should be unsurprising. The assumptions of the second synthesis, that new species require new genes and that complex species have many more genes than simple ones, are false. Whatever mathematical genetics measures, it does not measure the transmission of the chemical genes discovered by developmental biology.

Change of context sometimes may slow or accelerate the maturation of the organism and alter the phenotype radically. In organisms with significantly different stages in their development, as in tadpoles to frogs or caterpillars to butterflies, slowing or speeding maturation may result in the loss of a developmental stage. The evolution of vertebrates long posed a mystery because there seems no apparent connection between vertebrates and other multicellular organisms. The suspected evolution of vertebrates from the tunicate (sea squirt) is perhaps the most astounding example of change of phenotype through the loss of a developmental stage in all biological history.

The adult, reproductive tunicate is an invertebrate. It is immobile, attaching itself to stable objects like rocks, and it lives in colonies. It is utterly unlike a vertebrate. However, it has a larval stage. The larva is a bilaterian with a chord along its back. It moves. It resembles a sexually immature version of the urbilaterians. If sexual maturation accelerates enough, reproduction can occur in the larval stage. Then the adult stage is redundant and can be eliminated. The bilateral stage, then, is the only phenotype remaining. The animal now resembles a primitive vertebrate—or is one. In this way, biologists think, vertebrates evolved. The tunicate may well be our ancient ancestor.

As we know from anatomical, physiological, and chemical (DNA) evidence, we share a recent common ancestral population with the living apes. Although we do not look a lot like mature gorillas or even chimpanzees, we do resemble young apes. Biologists have long been aware of the resemblances. Our skulls are the shape of those of young apes, but not of mature ones. We have flat faces like young apes. New-born apes have little bodily hair except on their heads. We have little bodily hair except on our heads. Young apes lack the brow ridges so pronounced in their elders. We, too, lack brow ridges. In ape terms, we resemble sexualized children. Possibly, we are apes whose sexual maturation accelerated. Our immaturity (in ape terms) might explain our continuing curiosity, our creativity, and our flexibility in adapting to different cultures. Perhaps this is why we are so free, our behavior so underdetermined by our genes. As adults, we retain the curiosity, creativity, and flexibility of children.

Evo-devo has again altered the Darwinian phrase, descent with modification of traits. Under evo-devo, it has become descent with modification of development. Moreover, changes in development are controlled by embryo geography, by the cascades of events caused initially by core genes and their core processes, and by the context in which cells specialize. On the whole, changes in form or species do not require new genes. Insofar as this is true, evo-devo provides a simpler model than that advanced by the second synthesis.

Summary

All organisms share many genes and genetic processes. These core genes, clusters of genes, and developmental processes provide viability and stability for evolution to make successful changes. They lay the foundation on which variability occurs. They allow variation in forms and in functions that, nonetheless, let the organism live. Core processes keep it viable, even while changes occur.

Duplication supplies much of the variation. Genes may be duplicated, but so may clusters of genes, chromosomes, cells, organs, segments, and appendages. Duplication allows the original part to maintain its function while its twin specializes to become something different. The duplication of *Hox* genes resulted in increased complexity and the specialization of brains, organs, and bodily segments. The duplication of bodily segments and their appendages allowed insects to transform their new parts into different kinds of wings or legs or antenna, among other things. The duplication in snakes of spine and rib allowed them to eliminate legs and move about on their ribs. Organisms are modular. Modules occurring later in development can change, leaving the core modules stable. Modularity gives organisms the ability to evolve.

Evo-devo demonstrates that the embryo provides the key to evolution. Stability in embryo geography maintains the constancy of vital functions. Changes in embryo geography result in changes in animal form. Much of evolution consists in descent with modification of development—embryonic development. In this new model of the basis of evolution, on the whole new genes are unnecessary. Thus, as we peer back through time, we see organisms become simpler, yet contain many of the same genes found in more complex organisms.

Because almost everything in biology has exceptions, nearly everything in this chapter has exceptions.

7: SEEING THROUGH TIME

In biology, the kingdom is the largest, most fundamental division of organisms. Before the invention of microscopes, there were only two kingdoms, plants and animals. Now there are six. They are the archaebacteria, eubacteria, protists, fungi, animals, and plants. Both bacterial kingdoms are prokaryotes, that is, single cells without a nucleus that originated some 3.8 billion years ago. The protists are eukaryotes, cells with a nucleus, that evolved some two billion years ago as single cells. Fungi, animals, and plants are multicellular eukaryotes. They evolved about one billion years ago. All six kingdoms exist today. No one knows how the earliest cells originated.

Life began somehow

There is no clear definition of what life is, no explicit guideline to tell us whether something is alive or not. This constitutes one of the problems in establishing how life began. We know that the chemical elements from which life arose formed in the big bang and in the stars during preceding billions of years and that Earth developed from these elements some 4.5 billion years ago. The oldest rocks are about four billion years old, indicating that about half a billion years passed before the originally molten Earth cooled sufficiently to sustain stable formations, including the precursors to life. Like all dynamic systems, life requires energy to flow through interacting particles. However, even the source of the energy for the first life is uncertain. Before the discovery of archaebacteria, almost everyone assumed the sun provided the energy. But the archaebacteria thrive in extreme conditions, from acidic to salty and from boiling hot to below freezing temperatures. Most derive their energy from minerals far from sunlight. Because early Earth was hot, many biologists now think minerals and high heat may have driven the beginning of life rather than sunlight and moderate warmth.

Nor do biologists know the medium that enabled the precursors to life to form cells. Darwin suggested a warm pond where he knew bacteria thrive. A simple laboratory experiment by Harold Urey and Stanley Miller in 1953 supported Darwin's idea.

They ran electricity (energy) through a chemical atmosphere like that of early Earth, plus water, and created the precursors to life. Subsequent experiments altered the chemicals, yet produced similar results. Still, no one has created life this way.

Others think the medium might be clay, instead. Clays are rich in minerals, and crystallized minerals hold water and provide a rigidity that might have stabilized the first cellular membranes. Living organisms contain minerals and crystals similar to those in some clays.

However, the chemicals themselves may have originated in outer space. Complex organic chemicals form in space, and asteroids caught in Earth's gravity bring them to Earth. Many more asteroids struck the surface of the early Earth than strike it now.

So, what is life? Life on Earth is chemical, composed principally of organic compounds containing the element, carbon. From a biological point of view, life is anything capable of evolution. That is, living things must be able to multiply, inherit some traits, and vary down the generations. Life is subject to natural selection. Because the precursors to life lacked the features natural selection requires, the question, How did life evolve? is a nonsense question. Once life began, it evolved. But it did not start by evolving, because it lacked the requisite features. So knowledge of evolution cannot help solve the problem of life's origin.

As noted in the discussion of the bilaterians, biologists like to trace related life forms back to a common ancestor—actually a common ancestral population, not an individual. They coined the name LUCA for the *Last Universal Common Ancestors*—the initial life forms. Biologists know they had certain traits because all organisms possess them. All organisms display the same genetic code and share basic molecular and organizational characteristics. Because none of these characteristics is chemically necessary, they must have come from a common source, LUCA. However, Carl Woese, an outstanding expert on bacteria and early life, thinks LUCA may not have been organisms at all. That is, they may not have been whole cells, as all bacteria are. Rather LUCA may have been populations of precursors to cells that exchanged genes and evolved as a unit without having developed the cell membrane that separates cells.

The development of cell membranes is another major problem in uncovering the origin of life. A third is that of consistency. Life requires energy (metabolism) and heredity (genetics). These are two distinct requirements. How did they get together? Possibly, metabolism came first. Possibly heredity did. Perhaps, somehow, they developed together. Currently, biologists' best scenario is that the first life exploited RNA. DNA, the more complex hereditary material that existing organisms use, evolved later.

The search for the origin of life shows how science proceeds. At the moment, there are several contending theories for the origin of life. All have evidence supporting them. None has adequate negative evidence to refute it. Scientists search for both kinds of evidence. Someone may show that the formation of cellular membranes in water is impossible, refuting Darwin's original suggestion of a warm pond. Others may look for more evidence of the similarity between certain contents of living organisms and the mineral and crystalline content of clays. Some may seek further genetic similarity between ordinary living organisms and the archaebacteria that thrive in extreme conditions. Others look to outer space. Someday, perhaps, decisive evidence may refute one or more of these theories. Meanwhile, scientists pursue each theory hoping to find the true origin of life through one of them. Perhaps someone will develop a theory unimagined today that may synthesize the evidence, resulting in one grand synthetic theory, somewhat as Darwin did for evolution. No one yet knows the outcome.

However, in whatever way life began, it started long ago. It was microscopic. It was soft. It left no fossils. Moreover, once cells evolved, they likely ate it for lunch. No traces remain. Decisive evidence, if found, will be indirect at best. Even if we create life in the laboratory from chemicals, we probably will never be sure that life on Earth began exactly the same way.

For many centuries, the only known fossils were bones and shells. Bones and shells first appear in the Cambrian geological formation, beginning some 543 million years ago. Between about 525 and 505 million years ago, they become plentiful. This period is called the Cambrian explosion, because life seemed to burst forth suddenly then.

However, since the 1950s, fossil hunters have uncovered many Precambrian fossils and fossil traces. Now the first known fossils are bacteria, the prokaryotes. Among living prokaryotes are two groups that seem so different and so fundamental, Carl Woese divided them into kingdoms, the largest division of organisms. Eventually, other biologists agreed, so the oldest two kingdoms of life are the archaebacteria (bacteria mostly living in extreme conditions) and the eubacteria (bacteria living in temperate conditions). Originally, biologists considered archaebacteria the older of the two and named them archaea, for "at the origin." However, today biologists are uncertain which is older. In any case, sometime about 2.3 billion years ago, the archaebacteria and eubacteria split into two groups. Today eubacteria predominate.

The bacterial world was one of rapid change. Radiation bombarded Earth, causing many mutations. The bacteria had no repair mechanisms. Their life-style was simple: multiply, mutate, adapt, and exchange genes. Countless kinds must have arisen and gone extinct before archaebacteria and eubacteria evolved, stabilized, and thrived. And thrive they did! They were Earth's only living inhabitants for some two billion years.

Bacteria exchanged genes. They also ate each other. At times, the victims survived inside their predators. Margulis showed that the nucleus and other internal organs of eukaryotes evolved in this manner, by symbiosis, about two billion years ago. Some interior bacteria became chloroplasts that convert sunlight directly into energy. Cells with chloroplasts would later evolve by another act of symbiosis into plants. Others became the organelles in animal cells. Animals thrive only indirectly on sunlight. Their foods are plants and other animals. The symbiosis that produced eukaryotes required that the interior bacteria transfer their genes to the host nucleus. The nucleus in turn sent its products back to the chloroplast or organelle. Cells evolved hereditary abilities independent of their genotypes, through epigenesis. Epigenetic information is nowhere in the DNA.

The most creative period of cellular evolution ceased at the Cambrian. The core structures and processes were in place. Cells displayed metabolism, heredity, and organization. They engaged in sex. They communicated with one another and evolved the mechanisms that allowed them to stick together, enabling them to become multicellular. The three main sources for creating further evolutionary novelty were also in place: duplication with specialization, symbiosis, and epigenesis. The stabilization of core structures and processes constrained future evolution. Evolution had to maintain the core structures and processes. It also had to continue to operate according to the constraints posed by physical and chemical laws. Therefore, it could not produce just anything. It had to create within certain channels.

The core structures and processes may be compared to a 52 card deck. No laws say there must be 52 cards. None say they must be numbered one to ten, followed by jacks, queens, kings, and aces. The deck probably took thousands of years and many detours to create. However, it is now in place. Using it, we can play multitudes of different games merely by changing the rules and the roles each card plays. After the Cambrian, the evolutionary deck was in place. Changes of rules about when and where genes are expressed performed a major role in evolution from then to the present. Thus, the Cambrian provides a pivot point in evolution, but not an explosion.

The Cambrian "explosion" ambled

The Cambrian Period in geological history extends from 543 to 495 million years ago. The so-called Cambrian "explosion" occurred during some 20 million years in the middle of the Cambrian. It marks the origin of shells, bony internal skeletons, and external skeletons. These are hard substances. Unlike cells, they fossilize easily. Thus, the old idea that life originated in the middle of the Cambrian is mistaken. What evolved then were living forms easily seen without modern technology. Even ancient peoples picked up fragments of ancient bones and shells and speculated about them. No one before the invention of the microscope could see microscopic life, and the microscope itself underwent many refinements before it helped scientists study microscopic fossils.

The fossils in the Cambrian represent all but one of the animal phyla existing today. They also represent phyla that went extinct. Since the 1950s, biologists have discovered many microscopic fossils and fossil traces from the Precambrian. Easiest to find are multicellular eukaryotes. Some display exotic body plans. Yet, many are identifiable as precursors of today's phyla. Two to evolve hard parts are there. They are the arthropods and the vertebrates. Their common ancestors are our old friends, the urbilaterians. The urbilaterians possessed the master genes found in both phyla: *Hox* genes for segmentation, *Hk2* for hearts, *Pax-6* for eyes, and *Dll* for appendages. Their evolution marked an important break from their radial ancestors, whose round bodies left them directionless. The urbilaterians had a direction. They moved forward. Forward movement placed mouth and eyes at the front where brains would eventually evolve. Forward movement also provided many occasions for the specialization of appendages. As bilateral animals grew larger, they required an internal cavity to provide oxygen for the interior. They needed an improved pump to enhance circulation. The pump eventually evolved into a true heart, which built on the master gene *Nk2*.

As highly segmented animals, the arthropods evolved easily. *Hox* genes moved forward in the embryo, creating new segments that could specialize their appendages. Something similar happened in the vertebrates. The 20 million years of the so-called Cambrian "explosion" turned out to be a typical time span for animal radiations. Mammals evolved rapidly after dinosaurs went extinct, acquiring most of their current body plans in about 10 to 15 million years. Birds did the same, once flight evolved. Indeed, rapid evolution of new species in a new niche followed by slower evolution over succeeding millions of years is typical rather than unusual. There is nothing extraordinary to explain about this period in the middle of the Cambrian.

Except this: Why now? The precursors had been in place for half a billion years. Why, between about 525 and 505 million years ago, did this radiation of animal forms occur when no new niches seem to have opened up? Every new form lived in the sea, after all, just as their precursors did.

Many old answers are wrong, for they were based on the mistaken notion that life started in the Cambrian. Three possibilities remain, and perhaps all three played a part. First, perhaps sexual reproduction was ineffective until the Cambrian. This idea is difficult to test because sexual reproduction leaves no fossils. Second is the known increase of oxygen in the atmosphere about this time. Cells absorb oxygen directly through their membranes, and tiny animals through their skins. However, as size increases, it becomes more difficult, and finally impossible, to acquire enough oxygen in this manner. Internal cavities and tubes become essential. They increased at this time as size increased. Third, predation increased. This might have initiated the increase in size as well as explaining the origin of protective shells.

Many biologists see the Precambrian period as an evolutionary race to multiply body plans, with gene mutation and new gene creation foremost. After the Cambrian, the race changed into an arms (limbs) race, with duplication and specialization predominant. Ecology begins to drive evolution and, with it, extinction. From the Cambrian until today, steady, background extinctions occur that are typical of species in every geological period. Most live a few million years and then disappear. These are biologically-driven events, with natural selection eliminating individuals and their species as competitive struggles alter.

However, mass extinctions also drive evolution. These are physical events rather than biological ones. They, instead of natural selection, destroy organisms. They have nothing to do with adaptation. They constitute accidents. Only the lucky survive.

The most famous of these extinctions occurred 65 million years ago at the end of the Cretaceous. It wiped out the dinosaurs. Today, we know the cause. An asteroid struck the Yucatan Peninsula in Central America, sending tsunamis across the land and throwing rocks, debris, and chemical pollutants high into the atmosphere. Cold and darkness resulted, killing both plants and animals. Some 65 percent of species went extinct. Their extinction opened new niches for the rise of mammals. That violent asteroid cleared an evolutionary path for us.

The causes of the other four mass extinctions are obscure. They seem to involve atmospheric changes. Volcanic eruptions may have been responsible or at least a factor. The biggest extinction is also the oldest and therefore the most difficult to explain. It occurred at the Permian boundary some 245 million years ago and drove 95 percent of species to extinction. It very nearly rang the death-knell of life on Earth. Next came the Ordovician, some 439 million years ago, with 85 percent of species lost. Earth became colder. As life recovered, recognizably modern forms evolved.

The Devonian extinction occurred some 367 million years ago and eliminated 83 percent of species. The Triassic fell 208 million years ago and killed about 80 percent of species. Each of these events opened up new niches for the lucky species that survived. Some of those species underwent rapid adaptive radiations. If none of these mass extinctions had occurred—if fewer or more had occurred—we might not be here.

Life on Earth exists as we find it today because mass extinctions opened new adaptive zones. Organisms driven by natural selection were able to exploit them, to adapt to them, and then radiate out to fill empty niches. The availability of new niches speeds evolution.

New niches promote evolution

The only living organisms on land 385 million years ago were plants. In contrast, the oceans accumulated a diversity of living organisms, among them vertebrate fish that were neckless, with fins, conical heads, and scales. Twenty million years later, vertebrate fossils appear on land. These creatures retained many characteristics of fish, but they possessed necks, flat heads, and four legs. Thus, a fossil prediction: about 375 million years ago, fish came to land; the transition animal(s) will be found in strata of that time.

Exploring the predicted strata, biologists have uncovered some of the transition animals. One, *Tiktaalik,* has a neck and a flat head with eyes on top. It also has scales and fins. Inside the fins are bones, and the bones resemble those of animals with limbs. All animals with limbs have one bone in the upper arm/leg, followed by two in the lower, followed by various arrangements of wrist bones, followed by digits (toes/fingers). On *Tiktaalik,* the arm/leg bones and wrist bones are inside the fin. There are no digits. The similarity of all animals with limbs suggests one common vertebrate ancestral population conquered the land. It consisted of jawed, lobe-finned fish much like *Tiktaalik.*

What drove fish out of the water is unclear. Partly, they began hunting in shallow seas where both swimming and walking proved productive. There, fish possessing fins with bones that could bear weight were successful. Partly, smallish fish suffered heavy predation. Other fossils of the period show large fish with huge teeth. To survive, the smaller fish either needed to acquire armor, to grow bigger, or to move. Land may have provided a welcome refuge for the smaller, defenseless fish.

These, of course, are vertebrates. Six other phyla managed to conquer land, leaving 28 to remain in the oceans. Why did only some make it to land, where in the early periods there were few or no predators? Possibly, the body plans that developed in the Cambrian 150 million years earlier constrained them. Land, after all, poses many challenges to water-dwelling creatures. They lose the buoyancy of water and must contend with gravity. They leave the protective coat of water and must develop some means to deal with air that dehydrates the body. They must obtain oxygen from the air rather than from water. Hearing is affected because sound travels differently in water than it does in air. Perhaps the old tinkerer, evolution, simply lacked the material in those phyla to make the required changes. Or perhaps pressure to escape from the water onto land was lacking. The fossils tell little.

We know the story of the water-to-land saga of the vertebrates fairly well now. Vertebrates have hard teeth and bones that fossilize well. For plants, with their soft and easily-decayed parts, we know far less. The first problem, surely, was drying out. Land plants evolved waxy coats on their leaves and invaded land some 450 million years ago. Later, some evolved the structures that allow them to bask high toward the sun, shading their ground-hugging competitors, and to raise water from their roots to their tops.

The insect story poses similar problems. Insects are small, with parts that typically fail to fossilize. They came to land some 380 million years ago as tiny, crawling creatures. About 200 million years ago, wings evolved from their gills. These stiff, rather awkward wings later were replaced by the complex insect wings we know. Duplication and specialization tell the story. As insects added segments, *Hox* genes were repressed on all but the second and third thoracic segments. There the flying and stabilizing wings of modern insects evolved.

Whatever the exact story, insects were wonderfully successful. Several million species exist today.

Being slower to breed, birds speciated less rapidly. Nonetheless, they, too, were successful, radiating into approximately ten thousand species at present. The earliest undisputed fossil is that of *Archaeopteryx,* dated to some 145 million years ago. Birds evolved from feathered dinosaurs, and their wings use most of the arm.

Land and air provided new zones of adaptation for previously water- and land-dwelling creatures. Their evolution in unoccupied zones was typical of organisms that invade empty niches. It was rapid, requiring only a few million years. After that, body plans stabilized or evolved slowly over millions of years. Patterns of gain and loss in the DNA of the evolved species show no overall plan. They indicate what Darwin and his followers long suspected: natural selection acts only locally in time and space. It is not an engineer, with foresight and plan. We have similar DNA patterns, too, that show our evolution was local and unplanned.

As of 2001, the Human Genome Project decoded the human genome, so we know a great deal about our own DNA. We can also compare our genome with that of other animals—chimpanzees, to whom we are closely related, and fruit flies and mice, favorite laboratory specimens. Moreover, we are large mammals, with strong bones and teeth that fossilize well. Such evidence speaks of our past.

As Darwin thought, we evolved in Africa. We are primates, an order that evolved some 100 million years ago. Our nearest living relative is the chimpanzee. Our common ancestral population with chimpanzees lived some six to seven million years ago. From then until now, some 15 to 20 hominid species evolved. All but our own went extinct. We can best chart our ancestry as flared out like a bush, with no central trunk. The ancestral charts of most organisms are similar. We had lots of hominid relatives, but they do not form a chain, link upon link, reaching steadily to us. To seek "upward" links from other animals to us is futile. The "missing link" between us and the chimpanzees is a fantasy that borrows outmoded concepts from the European Middle Ages.

Like other creatures, we are modular. We became *Homo sapiens* one module at a time. The first transition occurred about 4.4 million years ago. Our first hominid ancestors walked upright. This is the transition that marks our ancestors as biologically human. However, they retained many ape-like features in their upper bodies, arms, and heads, including massive teeth and ape-sized brains. After that, as brain size increased, four more major transitions occurred. Hominids invented tools about 2.5 million years ago. About a million years later, some were carefully crafting their tools. They left thousands of meticulously wrought hand axes. Half a million years ago, they tamed fire. Yet, only 50 thousand years ago do scientists find evidences of the modern human mind. From that period forward, our ancestors left exquisite cave paintings, carved jewelry, and elaborate burials. Our brain had reached its present size.

Our big brain is expensive. When we are infants, it uses about 60 percent of our energy. In adults, it takes about 20-25 percent. Its energy expenditure suggests its importance in our evolution.

Despite many early scenarios of human evolution driven by ecological factors like male hunting (scenarios imagined by men!), the consensus now is that our big brains result from our social interactions, rightly giving the female half of our species an important role in our evolution. Most monkeys and apes are highly social. We have common social ancestors. Therefore, we were highly social even before we became human. Social monkeys and apes can track genealogy: they know who in their group is related to whom. They also track reciprocity: they remember who groomed them longest, best, and most recently, and who shared meat. They return the favors.

Interestingly, the size of the primate neocortex is correlated with group size. The larger the group, the larger the neocortex. When our ancestors moved from the forest out onto the African savannas as the climate cooled, they needed to form larger groups. In the forest, they could escape predators by hiding among the foliage or scrambling up the nearest tree. On the savannas, predators could see them from afar, and there were few trees to climb. Moreover, the savanna predators were swift and aggressive. Group living provided mutual aid.

Evolutionary psychology looks back to our millions of years on the African savannas. It claims (rightly) that our big brains evolved there. Thus, it says, our brains—our emotions, our perceptions—are best suited for a savanna life-style. Interestingly, three million years ago, hominid groups had about 55 members. In us, the size has expanded to 150. Scientists speculate that the formation of larger groups lies behind the invention of language.

Not only is language one of our outstanding attributes, but it develops spontaneously in human groups that initially lack language. The deaf have developed several sign languages. People thrown together by circumstance who lack a common language develop a kind of pidgin. Without tutoring, their children create fully-developed languages.

Approximately 50 thousand years ago, the direct ancestors of modern humans inhabited Africa. DNA studies indicate that an evolutionary bottle-neck occurred when only a few thousand individuals survived a massive die-out whose causes are still unknown. Due to an evolutionary transition still obscure in its details, they evolved symbolic thinking. Not only did they have a fully-developed language, but they made symbolic art. Their offspring invented elaborate cultures. Somewhere, the invention of culture reaches back to our genes, but less directly than the constructions other animals make, like spiders' webs and birds' nests and beavers' dams. Their constructions are few and specific to their species. The constructions in our cultures are numerous and take an uncountable variety of forms from curious fashions to shelters of earth, bone, skin, cloth, stone, wood, metal, and concrete, all of a multitude of shapes. Although our basic responses may be direct products of our genes and development, our cultures are not. We all, even infants, respond to rhythm. However, playing the piano rarely develops spontaneously, as many a suffering student can attest.

In truth, despite all our advances of knowledge, we know very little about our own minds or how mind and culture interact. The neurologist Oliver Sacks tells of autistic twins who, with evident joy, verbally shared prime numbers with one another. This constituted practically their only communication with others. Clutching a book of prime numbers, he joined them and offered a new prime number. They accepted it with delight. No one taught them.

Stories abound now about autistic people, unable to communicate successfully, who can draw elaborate buildings from one glance or play complex compositions on the piano, which they have only heard once. Are those abilities and prime numbers somehow buried in our genes? Or do our brains act more like radio receivers, tapping into some sort of universal knowledge? Perhaps this knowledge is more available to the autistic than to those of us whose attention is absorbed by social relationships. Despite all our discoveries, we remain a mystery to ourselves.

Summary

Awareness of deep time grew slowly, beginning in the eighteenth century. Although by the turn of the twentieth century, scientists knew Earth was ancient, they only dated it with certainty in the middle of the century. It is about 4.5 billion years old. Life began on Earth as soon as Earth provided a relatively cool home, about 3.8 billion years ago. No one knows how it began. The earliest life known to us is already complex—single bacterial cells lacking a nucleus. By symbiosis, nucleated cells evolved about two billion years ago, to be followed in another billion years by multicellular organisms. Modern organisms contain ancient genes, dating to the earliest bacteria, and probably to LUCA, the unknown last universal common ancestors. It is as if evolution invented a deck of cards (genes) that have been used to develop thousands of games (species). Evolution is a tinkerer. It builds new organisms from old parts.

The wrongly named Cambrian "explosion" never exploded. When biologists knew less about the past, they thought life arose suddenly during a short period of several million years in the middle of the Cambrian. Now fossil finds in Precambrian rock and DNA discoveries push the origin of life three million years further back. Certainly, animal forms radiated rapidly during the Cambrian, but no more rapidly than mammals did as soon as the dinosaurs went extinct. Before the Cambrian, evolution seems to have engaged in a body-form race. Afterwards, body forms stabilized, and evolution created an arms (limbs) race that was ecologically driven. Five mass extinctions reset evolution's clock, as the unlucky went extinct from physical trauma, and new niches opened for the survivors.

New niches promote evolution. Lobe-finned, jawed fish found a new niche for vertebrates on land and radiated rapidly, eventually evolving into the amphibians, reptiles, and mammals we know so well. Later, some vertebrates and insects took to the air, another new niche whose opening promoted the evolution of thousands of new species. As the climate in Africa dried and cooled and rain forests shrank, our hominid ancestors found a new niche on the African savannas. There, they evolved an upright gait.

Later, as their brain size increased and their bodies became more like ours, they invented tools and tamed fire. About 50 thousand years ago, evidence of the modern human mind appears in artifacts that are carefully wrought, cave paintings with delicate animal depictions, and burials that show imagination and awe. The symbolic mind had evolved.

Darwin speculated that life evolved from a common ancestral population and that we evolved in Africa, but he could not provide detailed proof. We can. Scientists enlarged the proof of Darwin's theory as they filled in this knowledge. Watching evolution through deep time helps us increasingly appreciate Darwin's theory of evolution.

Science discovered most of this information during the last quarter century. The second and third syntheses matured earlier, before the rise of molecular biology. They accomplished the integration of genetics and social behavior into Darwinian theory before the unraveling of the genomes of human, fly, mouse, yeast, and bacteria. Their assumptions failed to include the close relatedness of all life through evolution's retention and reuse of ancient genes. They completely excluded knowledge of the embryo. With so much unanticipated knowledge in hand now, is a fourth synthesis necessary?

Biologists know far more about evolution through deep time than appears in this sketch.

8. CREATING A FOURTH SYNTHESIS

The New Synthesis developed in the 1930s and 1940s. Novel discoveries about genetics and development since then remain outside the Synthesis. Along with the Synthesis, they provide the ideas required to develop a new integration of biological information, a fourth synthesis. Biologists working on the cutting-edge of evo-devo have discussed this possibility and developed most of the suggestions made here. The arrangement of material and the suggestion that mathematical genetics be treated as an instrumentalist theory, however, are the author's own, as is the numbering of the syntheses. Many people working on the fourth synthesis would number it as the third, either disregarding sociobiology or placing it in the New (second) Synthesis. However, most would agree that the initial act must be to reassess the Synthesis itself.

The New Synthesis requires reassessment

The broad consensus among biologists today is that the New Synthesis, as far as it goes, has proven largely correct. Its original goal was to unite mathematical genetics with experimental and observational knowledge. Its architects particularly desired to connect mathematical genetics with the temporal aspects of evolution, including evolution through deep time and the classification of organisms based on their anatomy and physiology, which partly depends on their history. The seeming rift between the continuous variation among natural organisms and the discrete nature of Mendelian genetics needed bridging. The New Synthesis united all sufficiently that the large majority of biologists accepted it by 1947. It reaffirmed the key to Darwin's original model. It made natural selection an integral part of biology, as it had not been since Darwin.

The New Synthesis used Darwin's model. From him it rightly emphasized common descent with modification, adaptation through natural selection, and competition for survival. The original problem of variation seemed solved by mutation of individual genes and recombination of genes in sexual reproduction. Mathematical genetics addressed the problem of heredity by using mathematics to show how heredity of and changes in gene frequencies might produce evolution down the generations.

The discoveries in 1944 that DNA is the chemical of heredity and in 1953 that its structure is a double helix only seemed to support the Synthesis. They showed that mutation involves chemical changes. They explained how multiplication occurs by unzipping and copying the double helix. And they promoted experimentation on DNA's role in biological functions including development.

Mathematical genetics remained unaffected. This is unsurprising. Mathematical genetics filled an important philosophical function in biology. Everyone in the early twentieth century considered physics the foundational and exemplary science. Newton married physics with mathematics in 1687 when he explained motion by gravity. Mathematical physics was king! Many scientists considered biology its poor relative. Darwin's great works of 1859 *(On the Origin of Species)* and 1871 *(The Descent of Man)* were verbal. They lacked mathematical underpinning. Mathematical genetics supplied the missing figures. Moreover, its basic principle, the Hardy-Weinberg law of 1908, mirrored Newton's first law. Newton's first law states that a body will remain uniformly at rest or in motion unless an external force acts upon it. The Hardy-Weinberg law states that two alleles (forms of genes) will retain their frequency in populations down the generations unless external forces act upon them: immigration of new organisms into a population, mutation of genes, selection, non-random mating, or sampling errors (as present in a founder group isolated from its parent population).

The Synthesis ignored important biological functions that mathematics proved unable to capture. These included interactions among genes, sexual selection, embryology, and development. Mathematical genetics treated the organism as a black box.

Architects of the Synthesis complained. Ernst Mayr began referring to mathematical genetics as bean-bag genetics. It was a literal description. Text books in the 1940s and 1950s suggested laboratory exercises in which beans of several colors represented genes. Placed in a bag, they were then mixed and reassembled to represent the transmission of genes according to Mendel's law of independent assortment. But Mendel's law is false and was known to be false. The peas in his experiments had seven pairs of chromosomes, and Mendel studied seven traits, one from each chromosome. Scholarship on Mendel's work suggests he engaged in preliminary studies of the peas and rejected traits that failed to sort independently before conducting his famous experiments.

Biologists in the early twentieth century knew that independent assortment was false as a generalization about genes. One gene may lie behind more than one trait, or multiple gene interaction may result in only one trait. Another architect of the Synthesis, Julian Huxley, used gene interactions to close the rift between Mendel's discrete genes and organisms' continuous variation. The mathematically gifted Sewall Wright was especially sensitive to gene interaction. Verbally, he insisted upon it. Yet, in his calculations and accompanying graphs, he ignored it. It proved too complex to deal with mathematically. So was sexual selection. Moreover, biologists probably also ignored it because it seemed embarrassing. Discussion of sex was taboo. Worse, in Darwin's theory, females controlled males (oh, no!). Females influenced them to do tasteless things like sprout grandiose tails and wave them about in a sexually provocative manner. More than one biologist must have felt that discussing such behaviors would undermine biology's growing reputation as genuine science.

So, the New Synthesis failed to incorporate known facts and important ideas. In the advance of science, this happens often. Theories, even broad ones like the New Synthesis or Newton's theory of gravity, leave out awkward information. Newton thought his theory of planetary motion predicted that gravity would eventually pull the solar system into one big blob. Yet, Newton's model was so inclusive and so enlightening, science set the problem aside for a hundred years. The mathematics of the New Synthesis failed to deal with the organism. It treated it as a black box. However, in leaving it as a black box, the New Synthesis altered Darwin's model. Darwin made organisms central. They were the objects of selection. Their descent from common ancestors and the modification of their traits constituted evolution. The New Synthesis made genes central. They were selected. They descended from ancestral genes. They were modified. Many biologists who absorbed the Synthesis began to view evolution from the point of view of the gene. They constructed theories on this basis. Genes, however, have no point of view. They are chemicals.

Looking back from current knowledge, it is clear that a biological model treating the organism as a black box was inadequate. Failing to include the organism in the Synthesis forced mathematical genetics to become an instrumentalist theory rather than a realist one. The distinction between realist and instrumentalist theories is important. It is also technical. It goes back to the European Middle Ages. A little background may help.

Today, we think scientific theories must be descriptive. Even though they may present simplified models of the world, we think the models should describe how the real world works. This stance is realism. A theory that really describes real the world is a realist theory.

The European Middle Ages thought they already possessed a realistic description the world: the Bible and Aristotle together described it correctly. However, the description proved insufficient to do the work the medieval world required. It needed to predict accurately important dates like Easter. Prediction required mathematics. Mathematics provided an instrument for prediction. However, the predictions were insufficiently accurate. The Pope saw the need of a new calendar and asked the mathematicians of his day to develop a more accurate one. This required new mathematics. It turned out that the most accurate predictions used a model of the world that moved from Earth-centered to Sun-centered, which seemed to refute some passages in the Bible. This was not a problem. To the church, mathematics only provided an instrument for prediction. It never described the world. It was merely instrumental. Its models were merely instrumentalist. Galileo got in trouble not because his model was Sun-centered, but because he insisted it be treated realistically, as a real description of the world.

Modern science often treated imagined objects instrumentally, too. How could they be real when no one had observed them or discovered them experimentally? Their only role was to make the mathematics work. Among the imagined objects were atoms and, later, some constituents of atoms. Scientists agreed to treat them realistically only when they could observe them or manipulate them. Mendel's genes were imagined objects.

No one had seen them. Many biologists early in the twentieth century thought them instrumental rather than real. When chemists discovered DNA, biologists rejected it as the genetic material because it was too simple. It failed to fit their knowledge of inheritance and development. The discovery that DNA really is the material of inheritance in 1944 and the clarification of its structure in 1953 changed their minds. They now knew the chemistry and how it made mutation and inheritance work. However, in 1975 when E. O. Wilson published *Sociobiology,* he ignored the implications of the chemical gene for mathematical genetics. Wilson says explicitly that his goal is to combine mathematical genetics with the discoveries about the social behavior of organisms in their environments. The organism remains a black box. Sociobiology, however, drew social behavior and sexual selection fully into the Synthesis. This change was sufficiently transformative to constitute a third synthesis. Yet, this synthesis, too, depends on the genes of mathematical genetics.

Evolutionary biology's dependence on the bean-bag genetics of mathematical genetics dies hard. Biologists still seek one "gene for" their favorite trait. *Scientific American* for January 2009 (**300**, 1) claims to offer the most up-to-date information on evolution and its modern applications. However, out of 61 pages, it devotes only one sentence explicitly to evo-devo. The article titled "From Atoms to Traits" (pp. 52-59) emphasizes "key genes" that produce large changes. The very term in the title, "atoms," invokes bean-bag genetics. "Testing Natural Selection" (pp. 44-51) devotes 1½ pages, or 20 percent of its space, to the effects of a single gene on flower color and hence on pollinators. No doubt, occasionally a single gene does affect a single trait, as seen here. However, on the whole, that is not how genes function. Moreover, even that single gene lies sleeping, while proteins do the work. And the gene is affected by its context.

Evolutionary biology is only beginning to open the black box of the organism and create a model that includes the hard-earned knowledge of chemical genes and development. The biologists working on evo-devo are returning the organism to its significant place in evolutionary biology.

The organism returns

The organism has been a black box with two compartments. One contains the chemical gene. The other holds all the complexities of development.

As the articles in *Scientific American* indicate, letting go of "genes for" and embracing the chemical gene has proven difficult for mainstream evolutionary biology. Moreover, in article after biological article, the authors say genes provide a "blueprint" for organisms. This is false. At most, genes provide instructions for making proteins. Proteins, in turn, regulate genes. Proteins control regulatory genes, even the master genes. Structural genes, regulatory genes, and the governing regulatory genes, master genes, code for proteins. Feedback loops and cascades occur repeatedly. Genes never directly make an organism's traits.

Genes also behave differently in different contexts. It is as true to say that their contexts provide instructions for genes as it is to say genes provide instructions for contexts.

Moreover, there are other agents of heredity. Cells display heredity mechanisms independently of genes. Stem cells may develop into any kind of cell. But once the cell type is established, a heredity cell line develops. Liver cells reproduce liver cells; heart cells, heart cells; skin cells, skin cells. Heredity in cells depends on specific genes being evermore turned on or off in the daughter cells that were turned on or off in the parent. The state of the DNA in the cell, and the state of the cell itself, is inherited, not merely its genome.

The second and third syntheses thought sexual recombination the major cause of mutation and variation. Recombination, of course, applies only to a limited set of organisms, those that reproduce sexually. It now appears that duplication in all organisms, asexual and sexual, is the major cause of mutation and variation. The duplication itself constitutes a mutation. Natural selection constrains the original of the pair to continue its old functions. However, because the original provides these functions, the duplicate is free to mutate. Natural selection no longer constrains it. Its mutations may prove helpful, neutral, or harmful. They may result in new functions. Helpful new functions of duplicates result in the modularity of organisms we observe everywhere. They depend on the duplication of ancient genes.

The discovery of ancient genes amazed everyone. No one anticipated that genes dating to LUCA had copies in us. Indeed, no one tried to reconstruct LUCA because biologists thought their genomes differed from those of modern organisms. New discoveries proved this false. LUCA and their bacterial descendants provided many genes existing today in all organisms. This means there are no "genes for" bacteria or blue fin tuna or blues singers. We all share genes. All life is related. To capture the reality of the biological world, all life must be folded into a single evolutionary model.

Partly, life is related because of symbiosis, especially the symbiosis between the eubacteria and the archaebacteria that evolved into eukaryotes—and them into us. The New Synthesis emphasized diversity, the splitting of lineages. Lynn Margulis taught it otherwise. Evolution involves the combination and integration of lineages as well as their diversification. Genes get together. They cooperate. As Darwin understood, much evolution has nothing to do with head-to-head competition.

Darwin also recognized a powerful evolutionary force that may work against natural selection. This is sexual selection. Surely, natural selection would result in animals that are well-camouflaged. Predators will eat those they can see most easily, so brilliantly colored organisms could never evolve. Nonetheless, they did evolve. And the organisms stand out against the green-brown background of most vegetation. Females, apparently, preferred bright mates. As a result of female choice, males threw away their camouflage. Sociobiology drew sexual selection into a third synthesis.

In summary, the chemical gene does little. It never makes observed traits, but only the proteins that produce the trait, at best. Moreover, gene interaction is one of the major forces in evolution. Opening the compartment of the organism's black box containing the chemical gene exposes a world unlike the one either the second or third synthesis modeled.

Unlocking the developmental component of the black box is equally transformative. One of the great biological discoveries of the nineteenth century was that all organisms are composed of cells. Another was that all cells arise from other cells. Cells are alive. DNA is not. Cells reproduce themselves. DNA does not. It is passively copied in the context of the cell with elaborate aid from RNA and proteins. Cells provide metabolism, the energy of life. DNA does not. The origin of the cell is the central question of the origin of life.

Development in multicellular organisms begins with a single cell. As it divides, each new cell exists in the context of the other cells. The cells produce chemicals that provide the developing embryo with a geography—in bilateral animals, a head and tail, right and left, top and bottom. The chemicals and signals within the geographical boundaries influence the cells there to become particular types of cells. This process continues, forming more and more compartments until the shape of the organism appears. Further geographical divisions along with the exploratory behavior of later cells produce the whole organism in all its infinite complexity. Biologists do not yet understand the whole except in the simplest organisms.

The context in which a cell exists influences what the cell does. In development, context is all. Nor is all context within the embryo. At some time, the organism enters the ecosystem. There, the context of its environment influences it. Environment is why the arrow leaf plant displays two different phenotypes. Environment is why plants grow stunted: their soils are poor in nutrients or their air rich in pollutants. A theory of organic evolution that ignores the embryonic context and/or the ecosystem is an incomplete theory.

Mathematical genetics ignores the context. It models genes as if each gene stands alone, in no context at all. Given the new knowledge provided by developmental biology, the gene cannot be the unit of selection. As Darwin knew, natural selection sees organisms. Moreover, it sees the whole organism in all its integrated complexity. Thus, the deer population that evolves under predation pressure to run faster may end up with longer legs than its predecessor populations. However, the longer legs may be more brittle. Legs may break. Those deer with larger and/or stronger bones survive in this context, not those with the longest legs. Somewhere in development, cells shift context, proteins influence genes, genes are turned on or off. Perhaps they duplicate. Eventually, these changes result in changed appendages, and populations change. They evolve.

Knowledge of chemical genes and their role in development shows that the genes of mathematical genetics are imaginary. They are mere mathematical abstractions. Because the genes of mathematical genetics fail to model real, chemical genes, mathematical genetics is an instrumentalist theory, not a realist one. With this insight in hand, it is time to look forward to the construction of a fourth synthesis.

A fourth synthesis beckons

More than two-thirds of a century has passed since the completion of the New Synthesis. Biology made numerous discoveries during that time, most routine, some transformative. When we look at how scientific knowledge accumulates, it is clear that the second synthesis begins to look outdated. Darwin produced the first synthesis in biology. He offered a mechanism for evolutionary change, natural selection, that showed evolution possible. His theory united knowledge of fossils, anatomy, physiology, classification, embryology, and biogeography. It also explained them and made a variety of predictions. Yet, much was unknown. Especially troubling were the causes of heredity and variation, two of the three pillars of his theory.

New research programs began to make new discoveries. The rediscovery of Mendel's work in 1900 provided the needed stimulus. Within some 40 years, biologists had integrated Mendel's concept of genes with Darwin's understanding of evolution to produce a second synthesis. As with the first synthesis, not everything fit. The mathematics excluded gene interaction. Development was ignored. Sexual selection was avoided. Clearly, the New Synthesis was incomplete. Nonetheless, it was an important enlargement of Darwin's theory and increased its predictive power.

Sociobiology brought about a third synthesis. It expanded and reoriented the second synthesis when it included animal social behavior, based on mathematical genetics, into biology. And because animal social behavior is largely about reproduction, it pulled sexual selection into evolutionary biology, too. Both subjects had been major concerns for Darwin. He wrote a book on animal behavior, *The Expression of Emotions in Man and Animals* (1872), and another almost wholly on sexual selection, *The Descent of Man and Selection in Relation to Sex* (1871). Wilson's subtitle for *Sociobiology: The New Synthesis* was correct. Sociobiology provided a third synthesis. It fulfilled and deepened Darwin's theory and offered sound predictions. Still, the chemical gene and organic development remained outside either synthesis. Meanwhile, mathematical genetics shifted genes to the center of evolutionary biology, a shift sociobiology supported.

During the 35 years since Wilson announced sociobiology to the world, knowledge of genetics and development exploded. The new knowledge in genetics updated classification and the history of biology. All these need fitting into a fourth synthesis. A fourth synthesis might include five steps—the author's own list, although derived from current work in biology.

(1) The first step would be to preserve as much of the second and third syntheses as possible. For the most part, this is easy because they got almost everything right. They retained the essential features of Darwinism, including common descent with modification, competition, natural selection and sexual selection. Moreover, they explained the causes of variation and the causes and implications of heredity more fully and accurately than the original theory had. Finally, through the application of mathematics to a previously verbal discipline, they brought biology increasing powers of prediction and recognition as a genuine science.

The predictive power of mathematical genetics is remarkable, both in its original application to the transmission of genes across populations and down the generations and to its later application to animal behavior in sociobiology. Its simplified model works well. The problem with it is conceptual: its "genes" are not genes as we now understand them. By calling them "genes," it deceives. It leads biologists to notice traits, then to seek "genes for" them. It leads physicians to look for genetic diseases and to try gene therapy for them. But there are no "genes for" these things. Moreover, the deception the term provides is dangerous: people have died from attempts at gene therapy.

So it seems time to admit that mathematical genetics is an instrumentalist theory. Doing so might lead to replacing the term "genetics" with "organismic processes." Thus, the old mathematical genetics, which claimed to be about individual genes, would expand to include organismic processes. So-called "genetic" diseases would become diseases of organismic processes, which would lead physicians to look for multiple genes and a cascade of organismic activity rather than a single gene. Intervention might then occur at any step in the cascade rather than trying to alter genes, with all their potentially dangerous interactions and dependence on context.

Saving mathematical genetics in this manner would resemble what happened in science after Einstein proved Newton wrong. Newton's mathematics remains. Scientists commonly use them rather than Einstein's because they are simpler and usually accurate enough to do the job. Where their predictions fail, Einstein's more difficult work must be substituted. The mathematical genetics, like Newton's, is simple, useful, and predictive. Moreover, biology today finds itself in much the same quandary physics does. When dealing with quantum mechanics and string theory—with, tiny, fast-moving things—physicists find the mathematics impossible to do in detail. They estimate. An attempt to include the interactions of genes and organismic processes in mathematical genetics might end up in the same quandary, with the same inadequate solution. Perhaps it will best to retain the instrumentalist theory, while acknowledging it for what it is.

(2) The second step, surely, would be to return to Darwin's original insights. This would include at least four alterations in the second and third syntheses.

(a) Put the organism back at the center of evolution and natural selection. This will naturally pull the embryo and development into evolutionary theory where Darwin placed them, as is happening already in evo-devo.

(b) Broaden selection beyond the organism. Darwin originally did this when he suggested that the evolution of ants, bees, and wasps might be due to selection acting on the family. The third synthesis' introduction of kin selection takes care of this. Later, Darwin suggested that groups of human beings might be selected in addition to individuals and kinship clans. The concept of group selection without consideration of kinship has a checkered history. David Sloan Wilson is influential in dragging it back into biology from a long exile, and biologists are considering it seriously once more. The most "groupish" animals on Earth are human beings who share the same culture. Because of culture in us, selection appears to apply to the group, whether it does in other animals or not. This was Darwin's original idea. It has received some investigation, but merits more. Understanding it might help us grasp the causes of war and seize the means of peace.

(c) Emphasize the importance of the environment in the action of chemical genes. Everyone knows it is important. Nonetheless, discussions of genes and development often ignore it.

(d) Widen the meaning of homology, as has already occurred in evo-devo. Homology is similarity attributed to descent from a common origin. To Darwin, descent from a common ancestral population caused it. An example is the similar structures of the fins of fish, the limbs of vertebrates, and the wings of birds. Insect wings were excluded because insects and vertebrates appeared to lack a common ancestral population. Now, of course, we know there is one—and Darwin believed so. But the ancestor is very ancient. However, the same genes operate in insects and in us, producing segmentation and appendages. Homology should apply to similar genes inherited from a common ancestral population rather than merely to similar ancestral structures.

(3) The third step would be to include sudden events unknown in Darwin's day. The obvious events are the mass extinctions. Discussion of them would remind us of the unpredictability of evolution on Earth. It would help remove us from the pinnacle of evolution. It would assist biologists in discussing the current mass extinction, which we are causing.

Niles Eldredge and Stephen Jay Gould argued in 1977 that sudden speciation followed by long periods of stability marks the fossil record. They concluded that evolution proceeds by rapid speciation, followed by stability of species. Although announced as revelation, this was old news. As long ago as 1963, Ernst Mayr emphasized speciation by founder population, which leads to rapid speciation. Today, however, sudden speciation by micro mutations seems likely, as Jeffrey Schwartz has recently argued. This is of special interest because a central goal of the second synthesis was to integrate macro with micro evolution. Our new knowledge of regulatory genes, and especially their governing master genes, may provide a new means for that integration.

(4) Biology has rightly been intrigued by diversity. It wanted to know how so many species came to be and why they developed into a hierarchy of classification. Darwin successfully addressed this central question by showing how populations change over time. Change implied diversity, and diversity seemed to imply splitting: one species splits into two, then those species both split, making four, and so on exponentially until billions of species exist. However, Margulis demonstrated that species also unite: two species come together and function so jointly that they become one species. Moreover, she demonstrated that the unification of species plays an important role in evolution. This important evolutionary mechanism must be incorporated into any complete theory of evolution. It constitutes a fourth step.

(5) The fifth step merely provokes a question here, for research has only begun. Does the ability to evolve evolve? Put another way, does evolvability evolve? We know little about what happened two or three billion years ago when the only living organisms on Earth were bacteria, especially before the archaebacteria and eubacteria separated. But once eukaryotes evolved, evolvability seems to have accelerated. The eukaryotes, after all, evolved multicellular organisms, then numbers of phyla. Perhaps eukaryotes are hardly more various than prokaryotes, given the variety of bacteria. But complexity certainly increased. Is this a sign that evolvability increased with the eukaryotes? Has it increased since? Evolvability, too, is worth researching as part of a fourth synthesis.

Whether it previously increased or not, if we learn to control our own evolution, evolvability will evolve. But save that for the future.

Summary

Reassessment of the two syntheses of the twentieth century shows they got most things right. They borrowed Darwin's model: descent with modification, adaptation through natural selection, sexual selection, and competition. They explained variation and mutation more fully than Darwin had. And they introduced mathematics into biology. However, mathematical genetics modeled the organism as a black box and made abstract genes central. These were mistakes. They over-emphasized the importance of genes, made mathematical genetics into an instrumentalist theory rather than a realist one, and ignored important information about organisms.

When the organism returns to the center of evolutionary biology, the chemical gene and organic development enter evolutionary models in a major way. Chemical genes engage in interactions and symbiosis. They duplicate, providing variation for natural selection. Many are ancient. They date back to the first bacteria and even to the first organism. In development, they respond to context and also provide it, as embryos acquire a geography.

To develop a fourth synthesis requires retaining as much as possible of the second and third syntheses. However, mathematical genetics needs reconceptualizing. It is an instrumentalist theory. Its genes are abstractions. A more complete synthesis must return to Darwin's emphasis on the organism, group selection, and the environment. It needs to widen the definition of homology to include homologous genes, consider sudden speciation, and incorporate the melding of two or more species into one. Finally, biologists need to define evolvability and research it to see whether it has increased as evolutionary history has expanded. If so, evolvability should be included in a fourth synthesis.

A complete fourth synthesis would require far more detail than conveyed here, and probably additional steps as well.

CONCLUSION

Darwin's publication of *On the Origin of Species* in 1859 marks the beginning of modern biology. His work accomplished four fundamental things. First, it offered a mechanism to drive evolution. Second, it formed a synthesis of the knowledge of organisms and fossils known at the time, while transforming that knowledge. Third, it made testable predictions. And fourth, it inspired research that is still continuing. The four taken together constitute signs of a brilliant scientific theory.

To the late nineteenth century, Darwin's mechanism for the transformation of species—natural selection—appeared too random, too dependent on chance occurrences. On the whole, almost everyone ignored it or rejected it outright. In contrast, most people accepted evolution—descent with modification of traits. Evolution revolutionized studies of biology. Classification and comparative anatomy became searches for common ancestors. Fossil discoveries provided evidence to reconstruct the history of life on Earth. Biogeography began to study the movements and transformations of species and became aware of the importance of founder populations. Each of these refashioned studies highlighted the importance of heredity. Heredity, however, stood as an unsolved problem. Yes, some traits were plainly inherited, but how?

The rediscovery of Mendelian genetics in 1900 addressed the problem through mathematical statistics. At the same time, it seemed to challenge the continuity of species. Darwin's synthesis threatened to fall apart. Aware of the problem, biologists worked together to form a New Synthesis by 1947. It united mathematical genetics with descent with modification and affirmed the importance of Darwin's mechanism of natural selection. This was truly an important synthesis. Yet, unlike Darwin, it emphasized the modification of genes rather than traits. Moreover, the New Synthesis, although comprehensive, remained incomplete. It excluded significant elements of Darwin's synthesis.

Sociobiology formed a third synthesis by pulling two of Darwin's original insights into the New Synthesis while enriching both of them. It emphasized and explained the importance of relatives in evolution while, for the first time since Darwin, including sexual selection. It increased the predictive power of the theory of evolution. Embryology, transformed in the late twentieth century into the study of development, remained the missing piece torn from Darwin's original synthesis.

The problem with development all along had been the difficulty of studying it. Development involves chemical genes that are both elusive and microscopic. Moreover, biologists must investigate their role in organisms that are alive. Chopping organisms into their component parts, a familiar technique in biology, proved useless. Biologists initiated knowledgeable studies of chemical genes only about 50 years ago. Only recently did technical proficiency become adequate for detailed examination of the inner growth of organisms. The new science of evo-devo—evolutionary developmental biology—is beginning to pull development into a renewed Darwinian synthesis.

This book offers at least three insights into the growth of our knowledge of evolution. First, Darwin's original work fulfills the requirements of good science. Evolution by natural selection constitutes a major scientific theory, one so important that thinking in biology can hardly be done today without it. Second, Darwin's work inspired and transformed research in many fields of biology. In the end, the new discoveries have vindicated Darwin's insights, expanded our knowledge of them by adding thousands of details, and enhanced our predictive power. Third, Darwin's original synthesis was so inclusive that new information seemed to threaten it. However, rather than tearing Darwin's synthesis apart, the new information expanded it by adding factual and theoretical knowledge. A grand synthesis embracing all contemporary information in biology remains incomplete. Far from being a criticism, such incompleteness is a mark of good science. Good science grows over time, raising new questions, inspiring further study, pushing the boundaries of knowledge ever outward. Darwin's theory exhibits its strength by its comprehensiveness. It has grown continually for more than a century and a half rather than suffering replacement by a more inclusive theory. This is a remarkable achievement.

Because almost everything in biology has exceptions, nearly everything in this book has exceptions. Biologists also know volumes more than is explained here.

ACKNOWLEDGMENTS

My thanks to the non-biologists who read and commented on the manuscript for clarity: Scott Fennessey, Lynette Friesen, Larry Hastings, Joan Jones, and Sara Sherrard. And gratitude to Marc W. Kirschner, an expert on evo-devo, for his helpful comments on chapter 6, "Appreciating Development."

Many people have contributed to the insights in this book from Peter Heath, who introduced me to philosophy of biology during my Masters program, to Michael Ruse whose fame lured me to the University of Guelph for my Ph.D., to Tom Settle, my thesis advisor, and Philip Kitcher, the external reader for my thesis. Special thanks to Ernst Mayr for writing so clearly and extensively on evolutionary biology. He has been my constant mentor-in-print.

Without these people, those mentioned in the "Authors and Works Cited," and many other authors, this book would never have been written. All errors, nonetheless, are my own.

GLOSSARY

Altruism: In biology, this is a technical term, which has nothing to do with motives. It is the sacrificial behavior of an organism that reduces its own chances of survival and/or reproduction while enhancing the survival and/or reproduction of another organism.

Biogeography: The study of the geographical distribution of living organisms and fossils across the globe.

Blending inheritance: The false nineteenth century belief that characteristics inherited from two parents merged together rather than being inherited separately, one from each parent.

Cambrian "explosion": The mistaken idea that life began suddenly during the middle of the Cambrian Period, between 525 and 505 million years ago.

Coevolution: The mutually interacting evolution of two separate species.

Determinism: The idea that all events, including psychological events, are caused by preceding events or natural laws.

Epigenesis: Inheritance outside the genes that does not involve changes in DNA sequence; development after the initial action of the genes.

Eukaryotes: Cells with a nucleus.

Evolution: Changes in populations of organisms.

Evolutionary psychology: Studies of human psychology based on the belief that genes from our ancestors' evolution on the African savannas—and perhaps the genes from their ancestors—influence our behavior today.

Founder population: A small population of a species separated from its larger species whose descendants evolve into a new species.

Genes: Inactive chemicals that constitute the units of heredity and provide templates for the formation of proteins.

Gene pool: The complete genetic composition of an interbreeding population.

Genotype: The genetic makeup of an organism.

Homology: Similarity among organisms (and, now, among genes) attributed to descent from a common origin.

Inclusive fitness: Emphasizes the role of genetic relatedness (relatives) in evolution.

Inheritance of acquired characteristics: The false nineteenth century belief that offspring inherit characteristics developed by the activities of their parents.

Instrumentalist: A theory or object thought merely to be useful for mathematical prediction, with no part in the description of the physical world and no physical existence.

Kin selection: The idea that families constitute the units of selection, in addition to individual organisms.

LUCA: The Last Universal Common Ancestor(s): The population of organisms from which all present life descended.

Master genes: Genes that control other regulatory genes, often stimulating cascades of gene activity.

Mutation: Heredity change in a gene or chromosome configuration.

Mutualism: (also symbiosis) Species living together for their mutual benefit.

Natural selection: Darwin's greatest contribution to our understanding of evolution, the force in nature that causes some organisms to survive to reproduce, others to die before reproducing.

Neo-Darwinism: (also the New/Modern Synthesis) The merging of Darwinian evolution and Mendelian genetics between 1937 and 1947.

Nepotism: Showing undue favoritism toward relatives.

New Synthesis: (also neo-Darwinism, the Modern Synthesis) The merging of Darwinian evolution and Mendelian genetics between 1937 and 1947.

Organelles: Small "organs" inside eukaryotic cells, for example, mitochondria and chloroplasts, thought to derive from the initial symbiosis of prokaryotic cells.

Phenotype: The observable attributes of an organism produced by the interaction of genes and environments.

Population: An interbreeding group of organisms, usually a smaller unit than a species.

Population genetics: Statistical studies of the transmission of genes in changing populations.

Prokaryotes: Cells without a nucleus.

Realist: The idea that our theories and the entities they describe are truly present in physical nature.

Reciprocal altruism: see Reciprocity. Adding the term altruism to the term reciprocal has confused many biologists and their readers.

Reciprocity: The mutual exchange of goods and services, not considered altruistic.

Recombination: The formation of new gene combinations in offspring that do not occur in their parents.

Regulatory genes: Genes that control other genes.

Selfish: A misleading term applied to genes. Genes are chemicals. They have neither selves nor motives. They work together (unselfishly!) to produce organisms.

Sexual selection: Selection based on mate choice, usually by the female; also male-male competition over females.

Sociobiology: The study of animal social behavior, including human behavior, based on genetics.

Speciation: The division of one species into two or more species.

Symbiont: The technical term for organisms living in symbiosis, especially the smaller one.

Symbiosis: (also mutualism) Species living together for their mutual benefit.

Transmission genetics: Statistical studies of the transmission of genes down the generations.

Urbilaterians: The hypothetical first population of bilaterally segmented animals.

Virus: A complex molecule that is not alive but is a parasite that can engage in the activities of life—metabolization and reproduction—by exploiting the machinery of cells it invades.

AUTHORS AND WORKS CITED

Alcock, John, 2001. *Triumph of Sociobiology.* Oxford: Oxford University Press.

Buss, David, 1999. *Evolutionary Psychology: The New Science of the Mind.* Boston: Allyn and Bacon.

Carroll, Sean B., 2005. *Endless Forms Most Beautiful: The New Science of Evo-devo and the Making of the Animal Kingdom.* New York: W. W. Norton.

Carson, Hampton L., 1982. "Evolution of Drosophila on the Newer Hawaiian Volcanoes." *Heredity* **48**, 1, 3-25.

Crick, Francis, see Watson.

Darwin, Charles, 1964. *On the Origin of Species by Means of Natural Selection.* Cambridge, Mass.: Harvard University Press.

Darwin, Charles, 1965. *Expression of Emotions in Man and Animals.* Chicago: University of Chicago Press.

Darwin, Charles, 1981. *Descent of Man, and Selection in Relation to Sex.* Princeton, New Jersey: Princeton University Press.

Dobzhansky, Theodosius, 1982. *Genetics and the Origin of Species.* New York: Columbia University Press.

Eldredge, Niles and Stephen Jay Gould, 1977. "Punctuated Equilibria: An Alternative to Phyletic Gradualism" in T. J. M. Schopf (ed.), *Models in Paleobiology.* San Francisco: Freeman, Cooper, pp. 82-115.

Gerhart, John C., see Kirschner.

Gould, James L., 1977. *Sexual Selection: Mate Choice and Courtship in Nature.* New York: Scientific American Library.

Gould, Stephen Jay, see Eldredge.

Hamilton, William D., 1964. "The Genetical Evolution of Social Behaviour. I and II" in *Journal of Theoretical* Biology **7**, 1-51.

Hull, David L., 1974. *Philosophy of Biological Science.* Englewood Cliffs, New Jersey: Prentice-Hall.

Huxley, Julian, 1964. *Evolution: The Modern Synthesis.* New York: Science Edition.

Jacob, Francois, 1977. "Evolution and Tinkering," *Science* **196**, 1161-1167.

Kappeler, Peter M. and Carel P. van Schaik (eds.), 2004. *Sexual Selection in Primates: New and Comparative Studies.* New York: Cambridge University Press.

Kirschner, Marc W. and John C. Gerhart, 2005. *The Plausibility of Life: Resolving Darwin's Dilemma.* New Haven, Connecticut: Yale University Press.

Margulis, Lynn, 1967, see Sagan (Carl Sagan was her first husband, Thomas Margulis her second).

Margulis, Lynn, 1970. *The Origin of Eukaryotic Cells: Evidence and Research Implications for a Theory of the Origin of Microbial, Plant and Animal Cells on the Precambrian Earth.* New Haven: Yale University Press.

Mayr, Ernst, 1982a. *The Growth of Biological Thought: Diversity, Evolution, and Inheritance.* Cambridge, Mass.: Harvard University Press.

Mayr, Ernst, 1982b. *Systematics and the Origin of Species.* New York: Columbia University Press.

Mayr, Ernst, 1988. *Toward a New Philosophy of Biology: Observations of an Evolutionist.* Cambridge, Mass.: Belknap.

Miller, Stanley L., 1953. "Production of Amino Acids under Possible Primitive Earth Conditions." *Science* **117** (May), 528.

Miller, Stanley L. and Harold C. Urey, 1959. "Organic Compound Synthesis on Primitive Earth." *Science* **130** (July), 245.

Raff, Rudolf A., 1996. *The Shape of Life: Genes, Development, and the Evolution of Animal Form.* Chicago: University of Chicago Press.

Ruse, Michael M., 1973. *Philosophy of Biology.* Atlantic Highlands, New Jersey: Humanities Press.

Sacks, Oliver, 1985. "The Twins." *New York Review of Books* **32** (3) (February 28), 16-20.

Sagan, Lynn, 1967. "The Origin of Mitosing Eukaryotic Cells." *Journal of Theoretical Biology* **14** (3), 255-274.

Schwartz, Jeffrey H., 1999. *Sudden Origins: Fossils, Genes, and the Emergence of Species.* New York: John Wiley and Sons.

Scientific American, 2009. **300** (1) (January).

Simpson, George Gaylord, 1944. *Tempo and Mode in Evolution.* New York: Columbia University Press.

Smith, John Maynard, 1988. *Games, Sex, and Evolution.* New York: Wheatsheaf.

Sober, Elliott and David Sloan Wilson, 1998. *Unto Others: The Evolution and Psychology of Unselfish Behavior.* Cambridge, Mass.: Harvard University Press.

Trivers, Robert L., 1971. "The Evolution of Reciprocal Altruism." *The Quarterly Review of Biology* **46**, 35-57.

Urey, Harold, see Miller.

Watson, James D. and Francis H. Crick, 1953. "Molecular Structure of Nucleic Acids: A Structure for Deoxyribose Nucleic Acid." *Nature* **171** (April 25), 737-8.

Wilson, E. O., 1975. *Sociobiology: The New Synthesis.* Cambridge, Mass.: Harvard University Press.

Wilson, David Sloan, see Sober

Woese, Carl and G. Fox, 1977. "Phylogenetic Structure of the Prokaryotic Domain: The Primary Kingdoms."

Proceedings of the National Academy of Science U S A. **74** (11), 5088–90.

Woese, Carl, O. Kandler, and M. Wheelis, 1990. "Towards a Natural System of Organisms: Proposal for the Domains Archaea, Bacteria, and Eucarya." *Proceedings of the National Academy of Science U S A.* **87** (12), 4576–9.

INDEX

Made in the USA
Monee, IL
07 July 2026